CLIMATE CHANGE BAFFLES BRAINS

Climate Charlatans Commit Intellectual Fraud on Reason

L. Rowand Archer

ARCHWAY
PUBLISHING

Archway Publishing books may be ordered through booksellers or by contacting:

Archway Publishing
1663 Liberty Drive
Bloomington, IN 47403
www.archwaypublishing.com
1 (888) 242-5904

ISBN: 978-1-4808-8098-6 (sc)
ISBN: 978-1-4808-8099-3 (hc)
ISBN: 978-1-4808-8097-9 (e)

Library of Congress Control Number: 2019911381

Print information available on the last page.

Archway Publishing rev. date: 08/30/2019

Contents

Introduction

EVERY DAY, climate change alarmists are telling us that increased CO_2 levels in the earth's atmosphere due to the burning of fossil fuels is leading its inhabitants to a catastrophic ending. Despite the fact that not one of their doomsday climate prediction over the years has materialized, these alarmists continue making these predictions without empirical evidence to support these extreme claims.

As an example, Democratic candidates running in the US 2020 presidential race are making the outrageous claim that the world will come to an end in ten to twelve years if nothing is done to curb climate change, but they offer no material evidence to back up their nonsense.

To maintain this hoax, these alarmists have enlisted the liberal big media, which seems not to care about its integrity and doesn't really care about honesty and good reporting. Instead, they pursue ratings and social media followers with their latest outrageous attempts to advance the climate alarmists' fake narrative.

To counter Mother Nature's pragmatic reality, the climate alarmists' consensus science has now moved their doomsday scenarios out fifty to a hundred years, so there is no way to scientifically test their computer-modeled nonsense. The most depressing part of the whole climate change issue is that otherwise sane people seem invested in this catastrophic silliness.

This book is intended to provide a nontechnical understanding of climate skepticism as argued by knowledgeable authors in their fields who question the notion that humankind is the major influence on climate change. In fact, because of its complexity and the limited knowledge of the effects of the different forcing agents in the climate scientific community, these experts cited in the book readily admit that it is really easy to be wrong and that a lot more research is necessary before climate science can ever be considered settled. Because weather is familiar to all, it seems that everyone has a theory about what causes climate change, and that makes it difficult to argue rationally about the real science behind climate change.

For climate scientists, recording data of the climate system is much easier than figuring out what that data means in terms of cause and effect. Generally speaking, it's not the warming that is in dispute—it's the cause of the warming. To determine if temperatures are continuing to rise, scientists can see only warming (or cooling) in the rearview mirror, that is, by looking back in time.

In the longer term, say hundreds to thousands of years, there is considerable indirect, proxy evidence (not from thermometers) of warming and cooling. Since humankind can't be responsible for these early events, it opens up the possibility that some (or most) of the warming in the last fifty years has been caused naturally. While many geologists like to point to much larger temperature changes that are believed to have occurred over millions of years, real climate scientists are uncertain that this past climate history tells us anything useful for understanding how humans might influence climate on time scales of ten to a hundred years.

Skeptical climate scientists argue that the bulk of the proxy evidence supports the view that it was at least as warm during the Medieval Warm Period, from 950 to about 1300, as it is today but cooled during the Little Ice Age from about 1300 to about 1850. However, the current observations that recent tree ring data erroneously suggest cooling in the last fifty years when in fact there has been warming should be a warning flag about using tree-ring data for figuring out how warm or cool it was over the past 2,000 years. But without actual thermometer data, scientists will never know for sure.

Other climate issues that focus on the melting of Arctic sea ice and rising sea levels are difficult to gauge since there is little to no recorded historical data to compare it with. As an example, since there are relatively accurate satellite-based measurements of Arctic (and Antarctic) sea ice only since 1979, it is entirely possible that the late-summer Arctic Sea ice cover was just as low in the 1920s or 1930s, a period when Arctic thermometer data suggests it was just as warm as today.

However, it is the increase in atmospheric CO_2 from a preindustrial level of approximately 270 parts per million by volume (ppmv) to today's level of about 390 ppmv that serves to support the main argument made by climate alarmists for the cause of climate warming in the last fifty years.

As I argue in this book, most of the CO_2 in the atmosphere was there before humans ever started burning coal and driving SUVs, yet alarmists argue that adding more CO_2 should cause warming. However, it is the magnitude of this warming that provides the basis of the real question argued by climate scientists today. The difficulty in answering this question is that scientists have no way of determining what proportion of the warming is natural versus created by human activity.

This book outlines arguments being made as to the cause and effect of a warming Earth including the natural changes in the amount of sunlight being absorbed by Earth and its effect on natural changes in cloud cover. According to some climate scientists, long-term changes in atmospheric and oceanic flow patterns can cause approximately a 1 percent change in how much sunlight is let in by clouds to warm Earth. They argue that this minor change is all that is required to cause global warming or cooling.

Significantly, it is not that there is no evidence that nature is the cause; it's that there is a lack of sufficiently accurate measurements to determine if nature is the cause. This is a hugely important distinction, and one the public and policy makers have been misled on by climate alarmists, academia, and governments and their institutions such as the UN Intergovernmental Panel on Climate Change (UN IPCC) and National Oceanic and Atmospheric Administration (NOAA).

This book further argues that when researchers approach climate change, there is considerable evidence these scientists have preconceived notions often guiding them. It's not that their claim that humans caused global warming is somehow untenable or impossible; it's that they are compromised by political and financial pressures that result in these scientists almost totally ignoring alternative explanations for the observed climate change that is occurring. As well, to support their flimsy climate arguments, they promote unsubstantiated claims that ninety-seven percent of the world's scientists tell us climate change is urgent with no published names of the 97 percent of scientists or the 3 percent who disagree with them. They claim that unnamed experts are 90 percent or 95 percent certain that climate change is occurring due to human activity.

Despite the earth's climate being so complex, with scientific research often using it as an example of chaos theory, there has evolved a climate consensus among a small group of knowledgeable experts who promote a considerable element of groupthink, herd mentality, peer pressure, political pressure, support of certain energy policies, and the desire to save the earth whether it needs to be saved or not.

Unlike the global marching army of climate researchers that the UN IPCC has enlisted, climate skeptics do not walk in lockstep. These skeptics are willing to admit "something I know not what"; they do not mislead people with comments that equate daily extreme weather with an increase in CO_2 levels and lead the public to think that climate scientists have determined through extensive research into all the possibilities that daily weather and the long-term climate cannot be due to anything but CO_2.

As well, climate skeptics advance alternative explanations or hypotheses for climate variability based on the way the research community used to operate using the tried-and-proven peer-review processes before politics, policy outcomes, and billions of dollars got involved.

The science of climate change has produced the myth of settled science despite the fact that science is never settled. There is, however, an anticapitalist political agenda behind the claims of many climate alarmist scientists. That is, they claim Western societies must radically reduce their standards of living to prevent climate catastrophe. Politics and ideology, not science, promote this so-called scientific view.

1

Climate Alarmism Has Become the Duty of Charlatans

BRUCE WALKER wrote the article "Madness and Political Life" (Walker 2008) in which he cites George Orwell: "We have now sunk to a depth where the restatement of the obvious is the duty of intelligent men." Walker asks,

> How often do we have to return to the words of Orwell to make sense of the world today? There is madness in our lives, and it neither seeks nor wants a cure. Great nations are altering their economies, impoverishing their poorest subjects, condemning their progeny all in support of the lie of global warming.

Despite the fact there is actual agreement on there being cyclical climate change as part of the earth's over 600 million years of climate history as addressed in this book, our current understanding of why these cycles have occurred is in its infancy and is far from being settled science.

There is debate as to whether the earth's climate is still warming after the Little Ice Age, which ended around the 1840s, or if the earth's climate is beginning another cooling period with temperatures leveling or slightly cooling since 1998. To counter Mother Nature's pragmatic reality, the climate warming alarmists' consensus science has now moved its doomsday scenarios out fifty to a hundred years, so there is no way to scientifically test their computer-modeled nonsense. The most depressing part of all this is that otherwise sane people seem invested in this catastrophic silliness.

Walker further argues that universities have become havens of official intolerance and bigotry the likes of which free-market societies have seldom experienced. More galling are the facts that almost everyone encourages young adults to go to these colleges and that parents and students spend vast amounts of money for this education while even more billions of dollars for university education is funded by taxpayers. Yet university administrators and professors seem overwhelmingly hostile to free-market societies and their ability to create prosperity for their citizens, to Judeo-Christian moral traditions in guiding the development of fundamental Western freedoms for its citizens, to its democratic political systems, and even to honest science for such common issues as global warming or arguments on the utility of Darwin's theory of evolution.

Further, Walker suggests that,

> Universities have almost morphed into an Orwellian Ministry of Truth. The dangers of a monolithic and totalitarian system of education in a democracy are obvious to anyone but the willfully ignorant. We have sunk to a depth where restatement of the obvious is the duty of intelligent men.

Stated differently, in "Tell a Big Lie and Keep Repeating It," Norman Rogers argues, "If you want to tell a big lie, a good vehicle is 'science'" (Rogers 2017b). The general public is gullible to these scientists' doomsday predictions because science and scientists have a positive image.

However, Rogers argues this positive image belongs to the science of the past, before the entrepreneurial idea of inventing fake catastrophes to attract vast sums of government money; lies backed by billions of government dollars make it difficult for the truth to compete against. A trusted truth can come only from scientists not corrupted by money. As well, Rogers argues,

> The truth is outgunned by the government's financed propaganda mills. These promoters of fake catastrophe depict themselves as disinterested idealists while the promoters of the truth are depicted as servants of evil industries or as mentally disturbed crackpots.

In the US, climate skeptics can lose their jobs and almost certainly be vilified as incompetent. So far, skeptics won't go to prison or to asylums, but there are calls by climate alarmists to criminally prosecute (Barkoukis 2016) "climate deniers."

Rogers concedes that if obvious stupid lies such as the competitiveness of solar power gain popularity and as long as those with the potential to expose these frauds are afraid to speak out, it's hard to see how fake science can be stopped or slowed down.

From Rogers's perspective, scientists who openly oppose the climate alarmist industry are invariably retired or otherwise occupy impregnable positions that protect them from economic retaliation. Rogers cites a list of 1,000 such scientists (Morano 2010) that oppose global warming alarmism. In contrast, there are very few young and upcoming climate scientists (Anderegg et al. 2010) that oppose the industry. The reason according to Rogers is if a young scientist opposes fake science, the individual will be unable to remain in the scientific field and therefore not be able to get a PhD and certainly can't get a job. For Rogers, "an entire multibillion-dollar industry is dependent on the public credibility of fake science. Dissent is not tolerated."

Despite there being plenty of scientists for whom science trumps money, the important bureaucrat scientists who exercise power and run things in such a way that money trumps science by a mile.

The conflicts inherent in the present sociopolitical and scientific arena of climate discussions are nicely summarized in Shakespeare's *Hamlet*, which pondered the eternal conundrum of competing choices: "Aye, there's the rub."

Rogers writes,

> As the Italian philosopher Wilfredo Pareto [Rogers 2015a] pointed out, people form their opinions based on passion. Resort to logic and data is basically window dressing meant to support their previously adopted opinions. That's why it is so difficult to make ideological conversions by means of logical argument.

Pareto's observation helps explain the stubborn attachment to the carbon dioxide dogma by climate alarmists. In "Climate Change Hysteria and the Madness of Crowds," Charles Battig expands on this observation by noting that after years of relentless doomsday prognostications by a variety of public voices spanning the political-scientific spectrum, they have found their mark in a gullible and guilt-prone public (Battig 2014). Battig writes,

> There is a Medusa-like quality in the serpentine web of doomsday prophets including members of the Club of Rome [Kuenkel et al. 2018], Paul Ehrlich's "Population Bomb" [Ehrlich 1971a], and the former White House science advisor John Holdren. In the US, the late Rachel Carson [Edwards 1992] proclaimed DDT to be environmental enemy number one and inspired Al Gore to discover "Inconvenient Truths," later found to be not so truthful [Adams 2007]. ...The repeated refutations of "faulty" science and failed predictions of climate calamities have not deterred these marketers of doom. Cut the head off, yet it lives on.

Issues arising from these doomsday prophets as noted by Battig include sustainability, population control, and redistributive-based social justice that were offered as moral justifications for the one-world governance needed to solve one-world problems as posited by the UK's Barbara Ward (Reuters 1981). Answering these world problems, Battig argues that "the UN's global bureaucrats crafted the IPCC as the instrument by which life-sustaining carbon dioxide would be reinvented as the most dangerous threat to the world." Governments around the world are more certain than ever that the science is settled and that the global climate bears the human stain of excessive consumption of fossil fuels.

Battig further argues that the climate alarmists have morphed their strategy of fear into a self-hate/guilt propaganda tool.

Humans are carbon-based life forms intertwined in biological interdependence on green plant production of oxygen and consumption of CO_2. Thus, the guilt stage is set for humans to be declared a living source of this newly defined carbon pollution and therefore enemies of mother Earth.

Battig cites that according to the Club of Rome,

> The common enemy of humanity is man. In searching for a new enemy to unite us, we came up with the idea that pollution, the threat of global warming, water shortages, famine and the like would fit the bill. All these dangers are caused by human intervention, and it is only through changed attitudes and behavior that they can be overcome. The real enemy then, is humanity itself.

It is not enough to just advance valid scientific truths to change the public's understanding of their environment to counter this climate doomsday propaganda. Instead, it is necessary to provide an adequate response that changes the public's emotional concerns for such issues as clean air, clean energy, and a healthy environment for themselves and their children.

Acknowledging this dilemma, Tom Trinko in "The Corruption of Science" argues that "the success of science in curing diseases and providing the basis for technologies that enhance human life has led to people thinking that science has the answer to all our questions" (Trinko 2018).

Trinko further argues that the reality is that not only is science restricted to explaining how things work and hence totally unable to address issues of morality or the meaning of life, but also, science itself is nothing more than a process by which we can objectively determine how physical things work.

To support his position, Trinko paraphrases Richard Feynman (Trinko 2014),

> The process of science starts with someone coming up with an idea of how to explain things, then predicting what should happen under certain circumstances, and then comparing the idea's predictions with the results of experiments. If the predictions match the data, then we say the idea is correct. If they don't match, then the idea is wrong.

Trinko adds that it does not matter who supports the idea or who opposes it; the truth is established by repeatable objective experiments, not by personal authority.

For Trinko, modern scientists are human and suffer from all the defects any sinner suffers from—their desire for power and money and the belief that their ideology is correct. Because of these failings, Trinko believes that just like most everyone else, scientists can lie to advance their own agenda whether for more funding or a change in the government's policy on an issue. When misrepresenting science, this corruption by scientists attempts to leverage the good science has done to advance their own personal agendas.

Adding to this fundamental scientific process, John Kudla states, "If a hypothesis is proposed and predictions based on that hypothesis happen as predicted, the hypothesis becomes a theory. If not, the hypothesis is rejected." {Kudla 2018}

Not so with global warming. When global temperatures fail to meet the IPCC's model predictions, they simply move the prediction date out into the future all the while making it clear the global warming apocalypse is still coming.

One of the primary problems for the global warming alarmists with their failed apocalypse predictions as outlined in Adam Yoshida article "The Warmist's Dilemma" (Yoshida 2011) is that they tend to live in the world of "Wouldn't it be nice" and then attempt to argue that people who dissent from this view are just plain bad. Nowhere is that tendency better exemplified than in the battle over their science interpretation of climate warming.

Yoshida contrasts this thinking to "conservatives and libertarians who allow themselves to get bogged down in the debate over whether global warming is occurring and whether humans are responsible for it." For Yoshida, this line of thinking turns the discussion into a charged debate over science that has the debaters "fighting an inconclusive battle of attrition over technical questions that are barely understood by the overwhelming majority of debaters" (Yoshida 2011).

The only way to arrest this global warming in the minds of the climate alarmists is through enacting state controls that will radically reduce the carbon dioxide produced by businesses and individuals. Without thinking too hard, this means that emissions must be reduced to such a degree as to be significant on the global scale. For Yoshida, this means one of two things—everyone must reduce his or her emissions collectively or free-market counties in the West must reduce their emissions so much as to compensate for the lack of reductions in the Third World.

Yoshida questions whether China and India, the world's largest polluters, are willing to take the sort of economically restrictive measures necessary to bring about a meaningful reduction in their carbon emissions and whether the US and Europe should engage in some sort of environmentally friendly mass suicide that will not have a net decrease over the medium term without the cooperation of China and India.

Yoshida and other climate science thinkers agree that to reduce carbon dioxide emissions in the short term can be accomplished only through heavy taxation and regulation of private enterprise and homes or through the mandating of the use of expensive and inefficient 'green' technology.

Alternatively, as discussed in this book, the long-term solution is to allow the private sector to innovate to bring low-cost, workable, low, or zero-carbon technologies to the market. To expect governments to achieve any kind of workable result using state-funded projects doesn't seem likely based on their poor historical record of picking winners.

As argued throughout this book, we should develop an understanding of the actual history of climate change cycles, which have been evidenced through real science of the earth's past. Instead of throwing trillions of dollars down green sinkholes, government and industry should be using this time and money to improve the economy and develop useful technology in an organic fashion—creating the breathing room and the sort of open and flexible society that can respond to the challenges inevitable climate changes might represent in the future.

The global warming alarmists' hypotheses of anthropomorphic (man-made) global warming (AGW) being caused primarily by carbon dioxide emissions is challenged by Dennis Chamberland in "The Tyranny of Consensus" (Chamberland 2015). In it, he argues that "climate change is the single most complex scientific question of human history." He argues that the global warming alarmists have

> managed to morph politics and science together into a repulsive, philosophic monstrosity—half science and half religion—specifically designed to reduce multifaceted, chaos-based theory and its inherent, profound complexity to absurdly simple computer modeled abstractions.

According to Chamberland, the reason was "specifically so that billions of dollars in global taxes may be levied at the point of a gun against the specter of anthropogenic climate change."

For Chamberland, science is the first line of defense against encroachments of ignorance, superstition and error is its own base of scientists and technical field experts ... It is the task of every scientist to be on the alert for failures in basic philosophy and to defend the integrity of the scientific method when necessary ... Science is not built upon its aggregate hypotheses—but the hypotheses are built upon and supported by science.

He believes that reversing this simple tool of philosophic understanding always results in serious error. Further,

> When that base has been so dumbed down by the wholesale collapse of a fundamental philosophic education prior to the awarding of degrees, it is inevitable that the institution would eventually be overrun with devastating but telltale errors in its most elementary philosophic tenants.

Chamberland states,

> The task is made even more difficult by an across-the-board failure of ethics within the profession, created by the billions of research dollars poured into anthropogenic climate change by a government that is entirely biased against any approach, study or theory except the one championed and paid for, solely reflecting the government's predetermined, ethically conflicted, politically and economically motivated, self-serving theories. For a scientist whose professional standing, and in some cases tenure, is based on research funding and publications, it is nearly impossible not to accept the government grants and just take the money. With

this money comes the published reputation of a true believer, and thus professional security is enhanced by not taking a stand against the anthropogenic climate change religion—basic philosophy be damned! Therein lies the problem that is resulting in the ongoing collapse of science, increasingly subsumed by the government's unyielding enforcement of its official doctrinaire religion.

In the biggest full-frontal assault on scientific integrity, in a single sweep, the US government changed (increased) as cited in an online article at Real Science "Mind-Blowing Temperature Fraud At NOAA" (Feynman 2015),

> 15 years of recorded temperature measurements specifically to match its version of the state's faith. All the data was altered in a single day. Thus, when the data fails, the government simply changes the data set as a veneration of the king's faith.

After this known fraudulent data alteration occurred, there was very little outrage from the scientific community. It was and is simply accepted as business as usual in this brave new world of government-enforced scientific faith in which anyone who speaks against the creed is labeled, ostracized, marginalized, and defunded.

Adding to these concerns of fraudulent data manipulation, Norman Rogers, in "The Corruption of Science" (Rogers 2014), cites a quote from Eisenhower's farewell address (Eisenhower 1961) warning that a "scientific-technological elite" dependent on government money would exert undue influence on government policy. Rogers now believes that Eisenhower's prediction has come to fruition with scientific advice contaminated by political considerations aimed at protecting the interests of their version of science and its scientists. He believes that scientists exaggerate the importance of their own work. While the corruption of science does not necessarily mean lying or faking data, it does come mostly from the exaggeration of its danger or importance while other scientists remaining passive.

To address the latest global scientific scare, Rogers exposes the intellectual foundation for global warming as being based on the smoke-and-mirrors predictions of doom that rely on computer models of the earth's climate.

> The only reason to believe these complicated models is the professional judgment of the very climate scientists whose jobs and reputations depend on the believability of those computer models.

At the time Rogers published this article, Mother Nature was not cooperating with the climate warming alarmists' computer model version of global warming as it had actually stopped about fifteen years earlier (Rose 2013). As a result, the alarmists stopped talking about global warming and started talking about climate change and extreme weather (Rogers 2013a). From a propaganda point of view, it is easier for the alarmists to blame every flood, draught, hurricane, and forest fire on extreme weather caused by adding carbon dioxide to the atmosphere since focusing on more-recent extreme weather is most vivid in the memory of the person in the street for whom the idea that the weather is getting worse seems plausible.

Chamberland writes,

> The damage thus far inflicted to the scientific process and the profession's integrity by the religion of anthropogenic climate change and its relentless tyranny of consensus, is immense. Like Diogenes seeking about the wilderness for but one honest man, it is becoming rare to find a scientist who will refuse to be counted on the list of the faithful, or speak up as a courageous, honest skeptic, challenging dogma,

therefore fulfilling the duty of a true master of scientific philosophy.

Richard Lindzen is a pioneer in climate science and a professor at MIT. In his essay "Climate Science: Is it currently designed to answer questions?" (Lindzen 2008b), he traces the history of government support of scientific research and describes how scientific seekers of government money adopted the promotion of fear as an incentive to open the floodgates of government money.

The reason governments provide their support to these scientific scare tactic is that the scientists show in their research such issues as radon pollution, DDT, nuclear power, the ozone hole, species extinction, vaccines causing autism, power lines causing cancer, fracking, secondhand smoke, ocean acidification, global cooling, and the biggest and latest scare, global warming.

As for the fear mongering of catastrophic climate change, it is working well for many climate scientists. Lindzen describes how these activists gain influence and power in scientific organizations and suggests that when there is a conflict between climate computer models (climate warming hypothesis) and data, there is a great tendency by these activists to change the data rather than revise their climate warming hypothesis to protect their status, but that happens at the expense of their scientific integrity.

The end-of-the-world fear strategy being peddled by money-sucking climate alarmists has been aided by ratings-seeking, liberal big media that started with CBS's Walter Cronkite's 1972 predictions (Seymour 2015) of a "new ice age."

In the UK *Guardian* article "World has three years left to stop dangerous climate change, warns experts" (Harvey 2017b), Fiona Harvey reported on an article by the same title in the science journal *Nature* in which the letter's authors argue that "the next three years will be crucial." If the Paris Agreement measures are not enacted even without the US, the world "may be fatally wounded by negligence before 2020."

It is important to note that this doomsday prediction has as one of its authors a former UN climate chief and a member of the IPCC.

In "More Fake News From the Climate Change Warriors" (Joondeph 2017c), Brian Joondeph cites another well-referenced example of climate fear by the high and mighty climate warrior Al Gore, who predicted (Watts 2016) in 2006 that "unless drastic measures to reduce greenhouse gases are taken within the next 10 years (2016), the world will reach a point of no return." Fast forward to 2019 and one can go outside and know that the climate still has not changed, that daily weather continues to fool local weather sources, and that on-air weather forecasters are confidently getting the predicted forecast mostly wrong for the next three days.

However, it is the liberal big media that have given into the global warming alarmists' fear-mongering propaganda and willingly exaggerate its significance. Joondeph cites the bastion of fake news, the *New York Times*, which "breathlessly predicted [Fox 2014] the end of snow" in 2014. Quite certain of the browning of normally snow-covered mountains in winter, they asserted, "This is no longer a scientific debate. It is scientific fact."

Reality exposes fake news. Joondeph draws attention to Mother Nature and how she had something to say about the *NYT*'s dire prediction when on May 22, 2017, the *Denver Post* announced that Aspen Mountain reopened (TAT 2017) for Memorial Day. "Mother Nature cooperated to make skiing an option." The reason for reopening was eighteen inches of new snow the previous week closer in the year to summer than winter.

Joondeph reminds us how ABC News has its own history of prognosticating about climate change; he cites a 2008 news special called Earth 2100. In this special, ABC made bold predictions (Huston 2015) based on the supposedly settled science of global warming arguing that by 2015, New York City would be under several feet of water due to rising seas. Gasoline would be $9 per gallon and milk $13 per gallon. The world population would shrink from 7 billion to under 3 billion due to climate catastrophes including shortages of food and water. As it turns out, all fake news.

A *New York Times* article entitled "U.S. Climate Report Warns of Damaged Environment and Shrinking Economy" (Davenport 2018) by Carol Davenport and Kendra Pierre-Louis (November 23, 2018) provides an example of the collusion between governments and liberal big media just before the UN climate conference in Poland (COP 24), which was convened to create new rules for cutting carbon emissions. (Note: it's not carbon dioxide emissions anymore; it's carbon, a completely different element). The article stated,

> A major scientific report issued by 13 federal agencies on Friday presents the starkest warnings to date of the consequences of climate change for the United States, predicting that if significant steps are not taken to rein in global warming, the damage will knock as much as 10 percent off the size of the American economy by century's end.

However, as reported in early 2019, the US economy grew by 3.1 percent in 2018 and in the first quarter of 2019 by 3.2 percent. More fake news.

In "Things Your Professor Didn't Tell You About Climate Change" (Kudla 2018), John Kudla argues that these same climate warming alarmists claim,

> Unless the buildup of greenhouse gases is stopped, global temperatures will begin to rise exponentially, which will have terrible consequences (Rinkesh 2018), such as major flora and fauna extinctions, coastal inundation caused by melting ice caps, heatwaves, drought, famine, economic collapse, war, and the potential for human extinction.

Kudla concludes,

> [The] basis for many of these predictions are the reports
> issued by the United Nations Intergovernmental Panel
> on Climate Change (IPCC). One of the functions of
> the IPCC is to model the Earth's climate to predict
> changes in global temperature. Although the Earth is
> warming a bit, their models always seem to be more
> enthusiastic about warming (Michaels et al. 2015)
> than the Earth appears to be. In fact, a recent study
> (Cockburn 2017) from the UK suggests climate
> models factor in too much warming.

The 2018 World Economic Forum in Davos, Switzerland, provided another opportunity for those who believe in man-made climate
change to harangue the thinking public about the evils of greenhouse
gases amid warnings that the world would end in 2050 or 2100 or one
of these days when it gets warm enough.

Most striking about these annual spectacles of the world's wealthy
and privileged is their disembarking from their fuel-gulping private
jets and limousines and emerging from luxury hotel suites to proclaim
the world must cut back on the use of fossil fuels or question why the
world's common people do not feel as deeply or passionately about
climate change as they do.

Kudla accepts the idea that the present warming trend may be
linked to rising amounts of carbon dioxide; however, it is still not settled science because scientists are still arguing over surface temperature data including the way it is collected, adjusted (Hoven 2018),
and interpreted. Even when these scientists have agreed on the data
sources, there are still climate science issues—whether carbon dioxide
is affecting global temperatures as much as believed and if water vapor
(Ball 2015), which humans have no control over, really dominates the
greenhouse effect in Earth's atmosphere.

Kudla cites peer-reviewed climate research that supports the hypothesis that the sun has a larger effect on the Earth's climate than any other climate forcing agent. Small changes in solar insolation due to variations in the Sun's energy output or cyclical variations in the Earth's orbit, known as the Milankovitch Cycles (Milankovitch 2019), can make a big difference in the surface temperature. There is little disagreement that greenhouse gasses, ocean currents, volcanic eruptions, and many other things can also affect the climate, however the Sun is still the 800 lb. gorilla in the room.

Due to these factors and other natural climate oscillations, such as the El Niño Southern Oscillation (ENSO), the 11-year Sunspot Cycle, the Pacific Decadal Oscillation (PDO) (Spencer 2018), the Atlantic Multidecadal Oscillation (AMO), and perhaps the De Vries Solar Cycle, the world's climate continually changes.

This means that the present warming trend is more likely a natural climate oscillation unrelated to just carbon dioxide but more likely to a combination of all the forcing agents some of which are not yet known.

Climate science is also certain that sea levels were higher in the past than they are today. In their 2014 Climate Report (IPCC 2014), the IPCC claims, "Maximum global mean sea level during the last interglacial period (129,000 to 116,000 years ago) was, for several thousand years, at least 5 m higher than present."

So instead of chasing shiny objects being thrown by President Trump and promoting bogus stories involving the president, the liberal big media should be asking the climate alarmists to provide proof of their outrageous arguments used to support their climate change theories and to explain their never-ending failed predictions.

2

The Rise of Global Warming
Dogma and Its Key Players

In "Global warming as Faith" (Rogers 2013c), Norman Rogers claims that global warming is a scientific theory but is mostly about "faith," with that faith playing a bigger role in science than we care to admit. He says that it was allegedly well-meaning intellectuals of the 1930s who believed in and defended Stalin's Russia in the face of massive and accessible evidence that helped "scientific" communism give birth to a terroristic, totalitarian state. Rogers compares today's believers in global warming to the intellectuals in the 1930s who fiercely defended their wacky faith in the face of massive and contrary evidence being vested in a theory that was precious to them.

For Rogers, it is when an ideology is precious that those who believe in it become aggressively hostile toward those who don't believe in it. The global warmers do not have dungeons or Siberian labor camps, but senior alarmists such as James Hansen, a scientist and the most famous global warming preacher after Gore, stated in "Twenty years later: tipping points near on global warming" (Hansen 2008), "CEOs of fossil energy companies know what they are doing and are aware of long-term consequences of continued business as usual. In my opinion, these CEOs should be tried for high crimes against humanity and nature."

According to Rogers, Gore thinks that people who deny his faith are like people who think that the moon landing was staged in Hollywood. In other words, those who question the global warming faith are either criminals or crackpots.

Rogers continues by making the following stark comparison between the IPCC and its relationship to the believers in global warming and the Communist International or Comintern's relationship to the believers in communism. The Comintern held world congresses and acted to enforce ideological conformity. The IPCC has grandiose meetings with delegates from many countries and publishes propagandistic tomes on climate science. Like the central works forming the intellectual basis of Marxism, the IPCC's publications are dense, turgid, and full of dubious science. Rogers argues that many of their claims have been exposed as phony in *The Delinquent Teenager who was Mistaken for the World's Top Climate Expert* (LaFramboise 2011) by Donna LaFramboise.

Rogers argues that there is nothing new about governmental organizations and highly educated scientists becoming fanatical followers of a global warming ideology since there have been similar movements and ideologies in the past that captured the fancy of intellectuals, scientists, and governments many times previously.

To make his point, Rogers provides the following example. In the first half of the twentieth century, the eugenics movement in the US was extremely popular and was embraced by the people such as the presidents of Harvard and Stanford and funded by foundations such as the Rockefeller and Carnegie. The eugenics movement advocated forced sterilization and restriction on immigration by so-called inferior races. These ideas were widely implemented by laws.

Fast forward—we see that global warming and closely related ideas such as clean energy, sustainability, biological diversity, green political parties, nature worship, organic food, and others have currently taken root in the intellectual imaginations of the masses of unemployed and underemployed graduates thanks to the seeds planted by their liberal leftist college professors. In *Intellectuals: From Marx and Tolstoy to*

Sartre and Chomsky (Johnson 2007), Paul Johnson argues that many thinkers have suggested that intellectuals as a class feel underappreciated and thus are attracted to ideological groupthink such as climate change to give themselves a sense of personal importance and become agreeable members of this chatter class.

To dig deeper in this climate change groupthink, we must understand its origin. In *Global Warming – A case study in groupthink* (Booker 2018) published by the Global Warming Policy Foundation (GWPF), Christopher Booker explains how only forty-some years ago, a few strong personalities suddenly turned the groupthink from advancing a theory of global cooling to formulating and perpetuating the groupthink of climate science being one of global warming. This switch to the warming-climate theory was aided by key positions and with sympathetic lobbyist groups that overwhelmed politics during its formative years in the 1970s from its center in various UN agencies.

In "Global Warming: The Evolution of a Hoax" (Leuck 2018), Dale Leuck identifies the first of those personalities as the late Swedish meteorologist Prof. Bert Bolin (Bolin 2008), who believed that increasing atmospheric levels of carbon dioxide from industrialization would inevitably lead to global warming. Bolin presented his views in 1979 at a first-ever meeting of the World Climate Conference sponsored by the World Meteorological Organization (WMO), a 191-member-country agency of the UN headquartered in Geneva, Switzerland.

Leuck notes that six years later, Bolin presented a paper for a 1985 conference in Villach, Austria, in which he concluded that "human-induced climate change" called for urgent action at the "highest level." An attendee who became convinced of this assessment was Dr. John Houghton, a former professor of atmospheric physics at Oxford who had been head of the UK MET Office since 1983. He was the founder of the Hadley Centre in 1990 and the lead editor of the first three reports (1990, 1996, and 2001) of the IPCC. The IPCC was established in 1988 by two UN agencies, the WMO and the United Nations Environment Program (UNEP). The UNEP was founded in 1972 by

Maurice Strong, its first director, as a result of the UN Conference on the Human Environment held in June 1972.

Maurice Strong, a Canadian self-made billionaire in the energy business and once self-identified as a "socialist in ideology, a capitalist in methodology," influenced the Stockholm Conference through a 1971 report he commissioned on "the state of the planet," a summary of the findings of 152 "leading experts from 58 countries." Strong later cofounded the Chicago Climate Exchange.

Leuck notes that the IPCC originally represented thirty-four nations with Bolin as its first chairman and Houghton chairing Working Group I, which was charged with contributing the climate change section of a first assessment report. Houghton wrote the summary of that report, which cited computer models indicating that global temperatures would increase "up to 0.5 degrees Celsius" per decade despite only a 0.6 degree increase in the previous hundred years.

In contrast, the text of the long report was reserved and underscored underlying uncertainty. The summary was designed to raise concerns in anticipation of the 1992 Earth Summit in Rio being organized by Strong and adding to the political momentum; it was attended by 108 world leaders, some 20,000 delegates, and an additional roughly 20,000 climate activists.

For Leuck, the only real proof of the scientific warming theory was computer models programmed to assume that increasing carbon dioxide was the most important factor driving climate.

Objecting to the predetermined conclusions of this first assessment report was Dr. Richard Lindzen, the Alfred P. Sloan professor of meteorology at MIT. In his 1992 paper "Global Warming: The Origin and Nature of the Alleged Scientific Consensus" (Lindzen 1992a), Lindzen pointed out that their models ignored other important factors that would have had cooling effects, namely water vapor, cloud cover, and oceans. Lindzen also noted that global temperatures had risen in the 1920s and 1930s, when carbon dioxide emissions were comparatively low, but fell back between 1940 and the 1970s, when emissions were rising much more steeply. He also concluded that the models would have predicted a twentieth-century warming four times more than actually recorded.

Lindzen also focused on how an illusion of a consensus had been used to dominate public debate by marginalizing credible scientists who evaluated the same data and disagreed with the prevailing conclusions.

Lindzen described how influential interest groups had fervently joined the cause of global warming such as the Union of Concerned Scientists, Greenpeace, Friends of the Earth, and the World Wildlife Fund (itself funded by Maurice Strong). Frank Press, the president of the National Academy of Sciences (NAS), stated (Pielke 2010), "Overt advocacy ... tended to delegitimize ... independent advice on science and policy."

The 1996 controversy was soon overwhelmed by the 10,000-person Kyoto Conference the following year, seemingly an example of consensus designed to continue the political momentum and later preparation for the 2009 U.N. Climate Change (Copenhagen) Conference (IPCC 2009).

The political and economic policy basis for the nearly total acceptance of climate change caused by the atmospheric trace element carbon dioxide, the fundamental requirement for life on earth, is tackled by David Archibald in "Someone Send the Coal People the Memo" (Archibald 2018). He cites a 1899 book published by Johann von Bloch, *Die Zukunft des Krieges* (*The Future of War*, Creveld 1899) and its author, a nineteenth-century railway magnate who built the Warsaw-to-Moscow railway. In those days, the best and the brightest worked on optimizing the productivity of railroads through operational analysis. Bloch applied insights from managing railroads to theorizing about the conduct of war. His big insight, original at the time, was that wars would be won by the country with the biggest industrial output. This is the same as the Soviet military concept of the correlation of forces as described in Julian Lider's article "The Correlation of World Forces: the Soviet Concept" (Lider 1980).

Archibald notes that the Soviets started using Bloch's insight in a big way until the 1980s, when the Chernobyl nuclear plant blew up in April 1986. The nuclear contamination was equivalent to a nine-megaton ground burst and was a big disaster for a country with a low standard of living. At a meeting of Warsaw Pact leaders three months later as revealed in records after the Belin Wall fell, the Bulgarian prime minister posed the question, how can the Communist Bloc profit from the Chernobyl disaster?

According to Archibald, for decades leading up to the time of that meeting, the Soviets had financed their Campaign for Nuclear Disarmament; however, the Warsaw Pact meeting triggered a new campaign against nuclear power in the West. The new strategy was to demonize nuclear power and convince the public to accept closures after the Russian nuclear mishap. Thus, Angela Merkel, herself the fruit of a long-term campaign by East German intelligence, was about to close the German nuclear industry without discussion using the Fukushima mishap as the excuse.

Archibald argues that this nuclear example has led to the more recent Communist disinformation campaign, with global warming becoming the biggest and most successful campaign, to reduce Western industrial capacity. As evidence, Archibald notes that on June 24, 1988, self-confessed global warming scientist James Hansen told a congressional committee, "Global warming has begun." This testimony was transmitted live to the offices of the European Environment Bureau and funded by the European Union in Brussels. According to Archibald, the Europeans were included in Hansen's testimony because they also wanted to hobble US industry.

The global warming campaign gained momentum but hit a roadblock in the US Senate with Republican senators asking why US industry should be hobbled with restrictions when Chinese carbon dioxide emissions were going through the roof and did not have to do anything to hobble its industrial expansion. Archibald further argues that China has adopted Bloch's insight and is doing what it can to hobble industry in countries gullible enough to believe in global warming.

The US dodged a bullet when Trump declined to sign on to the Paris Agreement despite the urgings of globalists Mattis, Tillerson, and Kelly and others such as French president Macron. This stopped the global warming movement in its tracks.

Providing supporting evidence to Trump's actions, Anthony J. Sadar, in "Has Progressivism Ruined Environmental Science?" (Sadar 2011), argues that from the start of the modern environmental movement with the publication of Carson's *Silent Spring* (Carson 1962) followed by *The Population Bomb* (Ehrlick 1968) by Paul Ehrlich, the science of the environment became overly contentious.

Sadar argues that despite diversity of opinion and positions in the scientific community being desirable and largely advantageous to the advancement of the discipline, "what quickly developed was a progressive environmentalism that elevated nature back to the Gaian status of the ancients and established one viewpoint as dogma." Soon, conformance to this one holy vision became the mandate just as the progressive mantra to celebrate diversity is today. "Anyone opposed to the lofty goal imposed by progressive theology of protecting the earth at nearly any cost was increasingly targeted in very unscientific ways with ad hominem attacks, public ridicule, eventual limitation of government funding, and even ecoterrorism."

To make his point, Sadar refers to the Climategate scandal in November 2009 at the Climate Research Unit (CRU) at the University of East Anglia with the hacking of a computer server by an external hacker who copied thousands of emails and computer files and Climatic Research Unit documents, several weeks before the Copenhagen Summit on climate change. The story was first broke by climate change skeptics with columnist James Delingpole popularizing the term Climategate to describe the email hack.

Sadar notes that despite the incriminating evidence in these hacked emails that exposed a gang of climate scientists manipulating the international climate warming discussion in favor of global warming, one would have reasonably expected much would have been made of the scandal. Instead, both self-interest groups seemed to work overtime to minimize the damage from those exposed files with the liberal big media often seemingly downplaying the importance of the email content and referring to the files as stolen. Similarly, the scientific establishment circled the wagons and has generally responded to Climategate by vilifying anyone who wants a serious independent or judicial investigation of the matter.

Progressivism is not a bad ideology for science. A variety of perspectives is beneficial and even rejuvenating. However, progressivism is bad if it is the only accepted philosophy by which to propose and evaluate concepts. Yet in the environmental field including climate warming, that is arguably the case.

3

Socialism—The End Game of Climate Alarmism

In "The End Game of Climate Change: Socialism" (Hendrickson 2018), Mark Hendrickson asks his readers to pretend for a moment that they have never heard of global warming or climate change and to focus on the alarm sounded by a press release issued by the UN IPCC, "Summary for Policymakers of IPCC Special Report on Global Warming of 1.5°C Approved by Government" (IPCC 2018), on October 8, 2018.

Hendrickson notes this press release calls for "rapid and far-reaching" changes "in land, energy, industry, buildings, transports, and cities." Among other major changes the UN wishes to oversee, as Brian McNicoll of Accuracy in Media summarized in "Time: Global Warming Must Be Stopped in 12 Years No Matter What" (McNicoll 2018), is a massive understatement: "All we have to do is eat a third less meat, move into smaller houses, use mass transit, and switch entirely from fossil fuels to renewables."

First, Hendrickson argues that climate change isn't about the environment because "the environmentalist movement today is dominated by an ideology that seeks massive government power over every important sphere of human activity."

Further, environmentalists believe that individual liberty must be curtailed. Environmentalism is simply the latest iteration of those illiberal ideologies—fascism, socialism, and communism—that share the worldview that the world is a critically messed up place that can only be saved if governments strip away individual rights and compel the masses to obey a government-designed central plan.

Hendrickson has no doubt that the UN leans heavily toward socialism. Goal number 10 of the UN Agenda 2030 aims to "reduce inequality within and among countries" that will be possible "only ... if wealth is shared." The document goes on to advocate "making fundamental changes in the way that our societies produce and consume goods and services"—typical Communist top-down central economic planning.

Senior IPCC official Ottmar Edenhofer said, "One has to free oneself from the illusion that international climate policy is environmental policy. Instead, climate change policy is about how we redistribute ... the world's wealth." Hendrickson argues along with a climate change insider that the real agenda of climate change is wealth redistribution; climate change is merely a convenient pretext.

Introduced to the Green Movement in the 1970s, Henderson contributed modest donations to various environmentalist groups such as the Sierra Club, the Cousteau Society, and others. However, based on their publications, Henderson realized that his donations were being used not just for lobbying for clear air and water but also for promoting an entire slate of liberal leftist causes including abortion, labor union privileges, and other socialist causes.

When the Soviet Union imploded in the early 1990s and socialism was discredited globally, American socialist intellectuals and activists, instead acknowledging the failure of government control of economic activity, morphed their desire for bigger government into the green garb of environmentalism and gave rise to the term *watermelon*—green on the outside but dark pink within.

Hendrickson cites a quote of Mikhail Gorbachev, general secretary of the Communist Party, from a Moscow conference in 1990: "The threat of environmental crisis will be the international disaster key to unlock the new world order." Gorbachev explicitly called for an international green organization to further that agenda. Henderson notes that Gore reportedly attended that conference.

Hendrickson cites the late Natalie Grant, an expert on Soviet disinformation who wrote in a 1998 article "Green Cross: Gorbachev and Enviro-Communism" (Grant 1998), that Gorbachev's plan included having environmental scare stories concocted and disseminated by pro-Moscow sympathizers and "useful idiots" in academia, the sciences, and the press.

In US and Canadian politics today is abundant evidence of foreign government involvement in domestic antifracking and anti-pipeline campaigns. Their goal is to cripple US and Canadian production of fossil fuels so Russia in particular can gain market share. For Henderson, this is where the Russians and the US greens make common cause.

Hendrickson further cites Kenneth Stiles, a retired CIA officer cited in an article "CIA Veteran Sees Russian Connection to 2 Groups Opposing Fracking, Pipelines" (Mooney 2017) who has found a money trail leading from Russian energy interests through Bermuda to various US environmentalist groups. Stiles declared that "Without a doubt … environmental groups are … agents of influence to Moscow through [a] networking system of shell companies and foundations."

Hendrickson speculates that it was suspicious that the IPCC issued its October 2018 dire predictions just a month before the 2018 biennial elections in the United States. The IPCC wanted to help Democrats retake control of Congress in the 2018 midterm elections, which was partially successful with the Democrats regaining control of the House of Representatives. For Henderson, environmentalists are still seeking revenge on Trump for having the courage to withdraw from the so-called Paris Agreement, which was an attempt at a huge redistribute-the-wealth scheme.

Andrew Thomas expands on the socialists' incessant "long march through the culture" by observing in "It's All About System Change (to Socialism), Not Climate Change" (Thomas 2014) that "environmentalism in America has been completely co-opted by the International Socialist Organization (ISO) and the Communist Party USA. Climate alarmism is all about killing capitalism and replacing it with a liberal left-wing dictatorship." Climate change is the vehicle to rally useful climate idiots. "System change, not climate change" is the new rallying cry of the socialist revolution.

PJ Media's Zombie writes in the article "Climate Movement Drops Mask, Admits Communist Agenda" (Zombie 2014),

> Communists along with a few environmental groups staged a 'People's Climate Rally' in Oakland, California on Sunday, September 21, in conjunction with the larger 'People's Climate March' in New York City on the same day.

However, in "Capitalism Is Green(er)" (Williamson 2014), Kevin D. Williamson argues that capitalism is proven to be a better steward of the environment than are socialism and communism.

Under a system that imposed heavy government regimentation upon the economy, direct government ownership of the "commanding heights" of the economy (and the commanded heights, too), a socialist vision of property, etc., the environmental results were nothing short of catastrophic. Setting aside the direct human costs of socialist environmental policy in the twentieth century—the famines, the deformations, the horrific birth defects—socialism was a disaster from the purely environmental point of view, too.

Williamson is convincing in his description of some horrific examples of environmental disasters wrought by the leaders of socialist governments during the twentieth century.

Thomas argues that since capitalism is an economic system separate from government as opposed to socialism, there are checks and balances in place. Some of these come from government regulation (that ideally should be minimal) but are primarily derived from the forces of competition in the market. For Thomas, the primary problem with socialism is that once you have the regulators and the regulated working for the same people, mismanagement and corruption inevitably occur and become pervasive. Environmental nightmares naturally ensue.

To further make his point, Thomas had the misfortune to observe in Shenzhen, China, where communism is the steward of this environment, that the air was headache-inducing and nearly unbreathable. The river running through the city was black as tar and almost as viscous. Knowing that the results of liberal leftist efforts are almost always 180 degrees opposite of their stated intensions, this could be a glimpse into America's future if they are successful.

Thomas Lifson adds,

> It isn't just demonstrator rabble unmasking themselves. Too much fanfare from the liberal big media, Naomi Klein, who passes for a leading socialist intellectual, who published a book, *No Logo* (Klein 2002), that openly states capitalism must be destroyed in order to save the planet.

The following description comes from Naomi Klein's website, www.naomiklein.org.

> Forget everything you think you know about global warming. The really inconvenient truth is that it's not about carbon—it's about capitalism. The convenient truth is that we can seize this existential crisis to transform our failed economic system and build something radically better. In her most provocative book yet, Naomi Klein tackles the most profound threat humanity has ever faced: the war our economic model is waging against life on earth.

This warped view of free-market systems has been advocated by the liberal left since Marx started advocating for his economic and political nonsense; it keeps shouting that anybody who disagrees with its politics is nothing but a stooge for capitalists or corporations. And wherever the liberal left come to power, it uses these claims to squash any dissent.

In "Climate Alarmism and the Muzzling of Independent Science" (Halperin 2016), Ari Halperin provides the example before Gore's tenure as vice president of when the majority of scientists with some knowledge of the subject firmly rejected climate alarmism.

> During Gore's term in the White House, he and the academic liberals executed a quiet purge. They packed the scientific establishment with environmentalists, defunded inconvenient research fields, removed distinguished scientists, and bullied others into silence or equivocation. Huge budgets allocated to climate studies (even before Gore) produced hordes of worthless PhDs incapable of making a living outside of climate alarmism.

But a large segment of scientists (Halperin 2016) and professionals versed in science, who are independent in a free society and derive their income from private business, disagreed with the climate-alarmist manufactured science. This posed a problem for Gore and other climate alarmists. Despite their shouting that climate skeptics were stooges of oil companies, they left a deep imprint on the public mind. It did not gain much traction until an opportunity came in the 1990s.

For Halperin, it was the antismoking campaigns that became the Trojan horse used by the liberal left to legitimize the suppression of conservative and moderate voices in the name of fighting against corporations. Halperin argues these antismoking efforts started in the mainstream with the intent to improve public health but then morphed into a campaign against smokers and tobacco companies that soon became a money grab. The first and the third phases had bipartisan support.

In the process, the Clinton-Gore administration developed and spread a false narrative—"People died because tobacco companies lied"—thus justifying suppression of corporate speech. Remarkably, the tobacco industry was blamed for conducting scientific research and even for funding research by independent scientists. The industry and/ or the scientists were blamed even if the research found or confirmed the dangers of tobacco and even if the research was unrelated to the tobacco. The Tobacco Precedent (Defyccc et al. 2001) was born.

By using the Tobacco Precedent, Halperin argues the environmentalists began to intimidate and vilify anybody who opposed climatim even casually. This strategy worked.

> Within a few years, multi-issue think tanks and conservative nonprofits were forced either to abandon their defenses of climate skeptics or to lose financial support from private industry including fossil fuel companies. At the same time, the environmental groups and climate alarmists, who had many times more money than the climate skeptics even before 1999, saw their income multiplied. This is how the climate debate ended.

See more detail in Defunding of Climate Realists (Defyccc et al. 2018).

In a series of articles including "The UN Wants to be Our World Government By 2030" (Ludwig 2018a), E. Jeffrey Ludwig recounts an encounter he had when he was an undergraduate at University of Pennsylvania with the chairman of the chemistry department, Prof. Charles C. Price, who told Ludwig that he was president of the United World Federalists, an organization that believed in a one-world government that would grow out of the UN. That surprised Ludwig, who at the time believed that the UN was a benevolent organization dedicated to pressuring the world community in the direction of peace and to operating charitable programs to help the struggling, impoverished peoples of the world.

Ludwig provided more details on the origin of the United World

Federalists (UWF) in a second article, "The 'Soft' but Real UN World Government Plan" (Ludwig 2018b). Ludwig states that after World War II, a group met in 1947 under the auspices of the UWF in Montreux, Switzerland, to draft a comprehensive plan for a world government (Montreux document, WFM 2019). The group stated unequivocally that they considered the UN "powerless ... to stop the drift of war." They believed that the establishment of a world federal government would be the only way to bring peace to our world. And many of that number believed that eventually, their six principles could and would be incorporated into the UN, which could morph into such a government. The UWF is specifically determined to mobilize the peoples of the world "to transform the United Nations Organization into world federal government by increasing its authority and resources, and by amending its Charter."

Ludwig argues that although there was a socialistic thread in its founding document, the UN was formed based on a vision of human rights presented in the "Universal Declaration of Human Rights" (UDHR 2019), which placed the concept of rights at the forefront for the progress of the world body. Rights are the mainstay for uplifting human freedom and the dignity of the individual. The UDHR document followed many amazing documents that presented rights as the central concept of the post feudal world: the English Declaration (or Bill) of Rights of 1689, the US Declaration of Independence with its important and forceful assertion of inalienable natural rights, the powerful US Bill of Rights enacted in 1791, and the French Declaration of the Rights of Man and the Citizen (1789).

Ludwig notes that there are new rights introduced that as early as 1945 were pointing the way toward intervention by the UN in the daily lives of people throughout the world. Throughout the document, they assert the right to food, clothing, medical care, social services, unemployment and disability benefits, child care, and free education plus the right to "full development of the personality."

Fast-forward to 2015—Ludwig notes that the UN published a

document of almost 15,000 words entitled "Transforming Our World: the 2030 Agenda for Sustainable Development" (Sustainable 2015). This document is a stream of consciousness of pious platitudes about meeting seventeen sustainable development goals and thereby meeting the most pressing needs of humanity. Except for the rights mentioned briefly, the word *rights* has been supplanted by the word *needs* to which men and women, rich and poor, Western, African, Asian, European, and Latin American nations can subscribe. Five of the seventeen goals are about the environment, and environmental issues take center stage because the environment affects all—rich as well as poor, highly developed industrial cultures as well as subsistence agricultural cultures.

Ludwig argues that with their new 2015 sustainable development declaration, "the UN has taken a giant step toward the socialist global government that was only hinted at in their first organizing document."

For Ludwig, the earlier ideas and ideals of rights, freedom, equality, and justice are subsumed under the meeting of needs and an explicit environmentalism that emphasizes preventing the depletion of scarce planetary resources. Ludwig suggests the takeoff is the Marxist axiom that society should be organized around the idea of "From each according to his ability to each according to his needs."

A skeptical Ludwig observes that,

> this projected transformation detailing a new world order of environmental responsibility and a significant reduction of poverty and hunger never speaks to the practical dimension of vast manipulations of people by cynical leaders and ignorant bureaucrats who hold their positions through terrorism and bribery. They never discuss incompetence and corruption, twin brothers in the family of venality. The document portrays a sincere world in which all those in power want to help humanity despite the daily evidence of the selfishness, corruption, murderous intents, devilish manipulations, thefts, personal immoralities, hatreds,

and utter depravity of many governmental leaders in every country in the world and among the leaders of business as well.

Ludwig then asks,

Is not the Agenda for Sustainable Development itself one of those devilish manipulations?

To provide further evidence that climate change is being used to deflect attention from the UN's effort to establish a world government order, in "Global Warming and Cold War Thinking" (Hawkins 2010), William R. Hawkins argues, "China sees climate change as another opportunity to help topple the United States from global preeminence, which remains its primary strategic goal in world politics."

To support this observation, Hawkins cites a press conference (Jiechi 2010) on March 7, 2010, held by Chinese Foreign Minister Yang Jiechi at the Third Session of the Eleventh National People's Congress in Beijing. Mr. Jang presented a masterful example of the diplomatic art of using the language of international peace and cooperation while at the same time promising, "We will continue to firmly uphold China's sovereignty, security, and development interests and conduct all-round diplomacy," demonstrating that it is more important to watch what these countries actually do in the world arena, not what they say. Hawkins asks his readers to consider the following statement from the foreign minister.

The world is moving toward multi-polarity, multilateralism and greater democracy in international relations. At the same time, a series of global challenges including climate change, energy and resources security and public health security have become more acute. More and more countries have come to recognize that the Cold-War mentality and zero-sum game theory have become anachronistic.

Hawkins notes that the first sentence has been Beijing's central objective since the collapse of the Soviet Union—to overturn the unipolar "hegemony" of American world leadership. New rising powers such as China want peer status with the US.

Hawkins argues that,

> the one place where Beijing wants democracy to prevail is at the UN, where China has staked out its leadership of the developing countries…The UN is supposed to be a venue for peaceful negotiations and international cooperation, but it is in reality a political battleground where rival blocs struggle to control the course of world affairs and restructure the balance of global wealth and power.

Hawkins cites,

> the December 2009 Copenhagen United Nations Framework Convention on Climate Change (UNFCCC), where the delegation had called for the rich developed countries to cut their greenhouse gas emissions by 25 to 40 percent by 2020 from 1990 levels. Such a drastic measure would have locked the developed countries into a permanent recession. Meanwhile, the developing countries led by the BASIC (Brazil, South Africa, India, and China) coalition and the Group of 77 (chaired by China's ally Sudan) would not have had any mandated restrictions on their GHG emissions. These nations, which make up the vast majority of the world's states and people, refuse to have any limits placed on their sovereign right to economic growth…

> [And] on March 9, 2010, China sent a letter (Finamore 2010) to the UNFCCC confirming that it would

attach itself to the accord reached at the end of the conference. However, it was in this two-track formula, or in UN terminology, the principle of "common but differentiated responsibilities," that the United States under George W. Bush refused to participate. The Obama administration held to the negotiating position of the Bush administration. The US view is that there should be only one track.

For Hawkins, this conference and all the others hosted by the UN are never about the weather. It was always a battle to gain economic advantages in a world of cutthroat competition. Chinese Premier Wen Jaibao confirmed Beijing's hard-line stance in a speech at the end of the Copenhagen meeting.

> The principle of "common but differentiated responsibilities" represents the core and bedrock of international cooperation on climate change and it must never be compromised ... Developed countries must take the lead in making deep quantified emission cuts and provide financial and technological support to developing countries. This is an unshirkable moral responsibility as well as a legal obligation that they must fulfill.

Beijing was playing on the widespread notion in the environmentalist and global warming movement that the world has a fixed carrying capacity and is thus a zero-sum game when it comes to economic growth. One of the most direct statements of this view, which makes international confrontations inevitable, came from Nick Dearden, director of the Jubilee Debt Campaign (Jubilee 2019).

> The rich world has gobbled up far more than its fair share of the earth's atmosphere in order to develop.

In essence, industrialised countries colonised the
atmosphere, in the same way they did other resources.

Those rich countries now owe poorer countries a twofold climate
debt first for overusing the earth's capacity to absorb greenhouse gases
and thereby denying atmospheric space to those who need it most.
The second was for the destruction that those emissions are causing.
The solution: rich countries need to pay through redistributing a fairer
share of limited atmospheric space as well as helping poorer countries
adapt to the mess they find themselves in.

The global Green Movement supported the demands of BASIC at
the talks and criticized Obama for not accepting large mandated GHG
cuts for the US. The battle continued at the succeeding major meeting
of the UNFCCC.

Looking at the attitude of the BASIC countries as expressed in their
reports under the accord posted at the UN Environmental Programme
(UNEP) site reveals that the pursuit of national advantage is still their
goal, not saving the planet.

Beijing's statement to the UNEP declares, "This is a voluntary
action taken by the Chinese government based on its own national
conditions." The UNEP cites an assertion by Minister of State for
Environment and Forests Shri Jairam Ramesh.

> I have been saying time and again that India, of all the
> 192 countries in the world, owes a responsibility not to
> the world but to itself, to take climate change seriously.
> We are not doing the world a favour. Please forget
> Copenhagen; forget the UN. We have to do it in our
> own self-interest. Our future as a society is dependent
> on how we respond to the climate change challenge.

Devoid of spin, New Delhi's response will be to put Indian development first.

The one-worlders of the 1950s and early 1960s are now in the UN driver's seat, and they have made their move. The overlay of Marxist talk about meeting needs has moved to center stage. The UN has assigned itself a time frame of 2030 for moving forward in its plan for planetary hegemony.

4

UN IPCC's Role in Corrupting Climate Science

In 2016, Ari Halperin published the article "The Paradoxical Origin of Climate Alarmism" (Halperin 2016) he begins by mocking the global warming alarmists by citing the fact that there were three feet of snow on the streets of New York and Washington in February, which was more than these citizens normally could expect.

It is Halperin's opinion that Mother Nature continues to show the alarmists that they get everything wrong.

> Anthropogenic carbon release is not dangerous or even harmful, but extremely beneficial. 15 percent of the world's agricultural production is due to the increased concentration of carbon dioxide in the air ... The global mean temperature has not been increasing for 19 years, and the slight warming expected from the emission of the infra-red absorbing gases is expected to be beneficial in itself [while] ocean water is alkaline, not acidic.

Halperin provides a link to a short summary of scientific (Goldstein 2018) errors by the alarmists and their logical fallacies, economic delusions, civic blunders, and cries of world doom. He asks how it happened that such a worthless agenda became so powerful. He believes the alarmists' outrageous claims and absence of scientific support became its political strength and a powerful public environmental cause.

Halperin argues that normally, real political and social issues such as drug addiction, poverty, illness, and abortion allow people to express different views and take different sides. However,

> The issue that gave birth to climate alarmism is different: the alleged problem (possible harm or danger from carbon dioxide emissions) simply does not exist. Most people are not interested in imaginary problems, and quite a few scientists, statesmen, and journalists came to the correct conclusion that the carbon dioxide/global warming/climate change problem did not exist

As a result, they did not participate in research or discussion on this topic, which allowed the believers to monopolize the topic. Competent scientists simply passed on addressing the alleged climate problem because they were not convinced of its significance especially compared to other changes happening in the world. For Halperin, climate change is real. The climate change problem is not.

The history of the possibility of man-made global warming from the release of carbon dioxide as a potential concern was seriously addressed in the 1970s when the government at the time convened top scientists to consider this potential concern. As stated above, this issue was of concern to few real climate scientists; the committee members became an overrepresentation of concerned scientists but few pragmatic scientists in the research discussion. Halperin noted one result of this self-selection: in 1979, the Charney and MacDonald (JASON) committees (Oreskes 2006) apparently working independently of each other arrived at the same climate sensitivity value, almost double the number that Guy Callendar (Applegate 2013) calculated in 1938, yet Callendar's results appear to be more accurate. Despite the observed results from these different research groups, none sounded the alarm or recommended that the government take any action other than continue with more research.

In September 1981 and for no good reason as noted by Halperin, the US Congress authorized the Carbon Dioxide Assessment Committee (CDAC 1983) and tasked it with a comprehensive evaluation of the possible dangers from carbon dioxide release. By 1983, CDAC delivered what became known as Nierenberg Report that advised "concern, not panic" and rejected climate alarmism again.

In a guest essay, "Who unleashed Climatism?" (Goldstein 2016), Leo Goldstein further outlines his understanding of the circumstances involved in the creation of global warming alarmism. Goldstein explains that climate alarmism erupted following the 1988 Toronto Conference convened by the United Nations Environment Programme (UNEP) and the WMO together with the Canadian government. Goldstein notes,

> After the conference, at which scientists constituted less than 15 percent of the delegates, the organizers and environmental Non-Government Organizations (NGOs) simply declared their alarmist claims as the new "scientific consensus," and threatened or defamed everybody who disagreed. Since then "the Big Lie, created by the United Nations (UN) agencies and environmental NGOs, has been thriving for almost three decades!

Goldstein observed that the IPCC was planned by the UNEP and WMO even before the Toronto Conference, which already had their own parallel science to that of the infamous James Hansen testimony before the US Senate initiated by Senator Wirth, who later became president of the UN Foundation.

This was followed up by an international web of alarmist organizations formalizing itself as the Climate Action Network at a meeting in Hanover, Germany. Initially, its headquarters were in Washington, DC, closer to power and money, but then it moved to Beirut, Lebanon, farther from law enforcement.

Goldstein argues that there are two persons most responsible for unleashing climate alarmism on the world. The first was Mostafa Kamal Tolba (Egypt), who headed the UNEP for seventeen years, 1975–92. Tolba was a microbiologist and a cabinet member of Nasser's government in Egypt who had proven his hostility to America by driving a wedge between the US and its Latin American allies in the negotiations of the Montreal Protocol on the protection of the ozone layer (Agrawala 1997). Tolba was tasked with providing support to government agents in adopting their country's policies to support IPCC's objectives.

The second person was Maurice Strong (Canada), the first head of the UNEP, a UN undersecretary, the organizer of Rio 1992 Earth Summit, and a man with three passions in life: power, money, and his openly expressed desire to make America a protectorate of the UN (Strong, 2000, 34, 313, 322, 329–38). He was deeply involved at the time with the top circles of the Democratic Party including becoming a trustee of DNC all without being an American citizen.

Halperin further argues, in his article "The Paradoxical Origin of Climate Alarmism" (Halperin 2016), that

> unfortunately for the world, UN agencies and environmental NGOs (mostly of foreign origin) picked up the issue and started running with it, which was not hard, because the real scientists and those who cared to consult them were staying out of the topic, rather than opposing the alarmist agenda…From the website of the IPCC,

> one gets the impression of an open environment for peer review interaction by climate scientists in their pursuit of the truth on climate change…

As well,

> the IPCC website claims that it provides regular assessments of the scientific basis of climate change, its impacts and future risks, and options for adaptation and mitigation.

> And it states that it was created in 1988 with the objective to provide UN government members at all levels with scientific information that they can use to develop climate policies. The IPCC reports are also a key input into international climate change negotiations.

This is where it starts to get interesting.

It is the promoted objective of the IPCC, to allow thousands of people from all over the world

> to contribute to the work of the IPCC. For the assessment reports, IPCC scientists volunteer their time to assess the thousands of scientific papers published each year to provide a comprehensive summary of what is known about the drivers of climate change, its impacts and future risks, and how adaptation and mitigation can reduce those risks.

Further,

> they claim to agree that "an open and transparent review by experts and governments around the world is an essential part of the IPCC process, to ensure an objective and complete assessment and to reflect a diverse range of views and expertise.

Through its assessments, Helperin argues the IPCC identifies the strength of scientific agreement in different areas and indicates where further research is needed. The IPCC does not conduct its own research.

Helperin argues the First Assessment Report (FAR) (IPPC 1990), completed in 1990, played an important role in establishing the Intergovernmental Negotiating Committee for the UNFCCC which provides the overall policy framework for addressing the climate change issue. In its scientific findings the FAR concluded,

- Anthropogenic climate change will persist for many centuries.
- Further action is required to address remaining gaps in information and understanding.
- There is continuing imperative to communicate research advances in terms that are relevant to decision making.

The Second Assessment Report (SAR) (IPPC 1995), issued in 1995, provided key input to the negotiations which led to the adoption of the Kyoto Protocol to the UN Framework Convention on Climate Change (UNFCCC) in 1997. According to Helperin,

> One of its main conclusions was: "The balance of evidence suggests a discernible human influence on global climate."

The Third Assessment Report(s) (TAR) (IPPC 2001), completed in 2001,

Helperin notes it concluded in its first working group report that,

> In light of new evidence and taking into account the remaining uncertainties, most of the observed warming over the last 50 years is likely to have been due to the increase in greenhouse gas concentrations. Furthermore, it is very likely that the 20th Century

warming has contributed significantly to the observed
sea level rise ... and widespread loss of land ice.

To repeat one of the IPCC main objectives, "An open and transparent review by experts and governments around the world is an essential part of the IPCC process, to ensure an objective and complete
assessment and to reflect a diverse range of views and expertise." Very
admirable, right? Not so.

As Richard S. Lindzen wrote in 1992 (Lindzen 1992a), "As [were]
most scientists concerned with climate, I was eager to stay out of what
seemed like a public circus." Cherry-picking was made easier by the
huge amounts of money lavished on climate science starting in the
1980s.

After the Toronto Conference, the intimidation and persecution
of openly dissenting scientists started. This is from the same article by
Lindzen.

> But in the summer of 1988 Lester Lave, a professor
> of economics at Carnegie Mellon University, wrote
> to me about being dismissed from a Senate hearing
> for suggesting that the issue of global warming was
> scientifically controversial. I assured him that the
> issue was not only controversial but also unlikely.
> In the winter of 1989 Reginald Newell, a professor
> of meteorology at the Massachusetts Institute of
> Technology, lost National Science Foundation funding
> for data analyses that were failing to show net warming
> over the past century. Reviewers suggested that his
> results were dangerous to humanity.

Contrary to this IPPC objective, in "Bali Climate Conference
Ignores Dissenter" (Moran 2007), Rick Moran exposes the UN as
being disinterested in allowing for a free and open debate on global
warming at the climate change conference held December 3–14, 2007,

in Bali, Indonesia. At this conference, the organizers refused to give credentials to a group of prominent scientists from the International Climate Science Coalition (ICSC) who would have presented a dissenting view from that of the global warming alarmists. To support this claim, Moran provides the following at the time,

> The United Nations has rejected all attempts by a group of dissenting scientists seeking to present information at the climate change conference taking place in Bali, Indonesia.

> The International Climate Science Coalition (ICSC) has been denied the opportunity to present at panel discussions, side events, and exhibits; its members were denied press credentials. The group consists of distinguished scientists from Africa, Australia, India, New Zealand, the United Kingdom, and the United States.

> The scientists, citing pivotal evidence on climate change published in peer-reviewed journals, have expressed their opposition to the UN's alarmist theory of anthropogenic global warming. As the debate on man-made global warming has been heating up, the UN has tried to freeze out the scientists and new evidence, summarily dismissing them with the claim "the science is settled." James M. Taylor, senior fellow for the Heartland Institute, explained,

> It is not surprising the UN has completely rejected dissenting voices. They have been doing this for years. The censorship of scientists is necessary to promote their political agenda. After the science reversed on

the alarmist crowd, they claimed 'the debate is over' to serve their wealth redistribution agenda.

Dr. Vincent Gray, an expert reviewer on the IPCC's published works, has debunked many of the claims of the IPCC in the past and was part of the ICSC group denied access to the conference.

Despite the notion that true scientists welcome open debate about their findings, Moran concludes the fake scientists at this particular UN conference were more interested in a political agenda than they were in discussing the science involved in their findings. For them, the science of global warming was settled. This objective is present to this very day.

Back in 2008, in "Global Warming? Bring it On" (Young 2008) Gregory Young was disgusted by the argument propounded by the dubious United Nations' IPCC report on Anthropogenic (human-induced) Global Warming (AGW) [being] willfully fraudulent ... [the] report has been vigorously and critically undermined, scientifically denounced and found wanting from both notable scientists here and abroad.

Instead of a true and open discourse, Young observes the daily dribble from the liberal big media and various liberally usurped science journals dishonestly and falsely alleging a consensus when there is none.

Young challenges the UN IPCC and its politically selected 2,500 scientists of whom a core group of 600 exists and a relatively small number of mediocre scientists here and there across the American landscape who have suddenly found notoriety or grant money in the global warming cause with the over 31,000 legitimate and viable scientists. One who signed the American Petition Project declared that the global warming hypothesis on the internet, "Global Warming Petition

Project" (GWPP 2019), where the group openly refutes the UN's conclusions as cited in the statement text of the petition, was bogus.

> We urge the United States government to reject the
> global warming agreement that was written in Kyoto,
> Japan in December, 1997, and any other similar
> proposals. The proposed limits on greenhouse gases
> would harm the environment, hinder the advance of
> science and technology, and damage the health and
> welfare of mankind.

> There is no convincing scientific evidence that
> human release of carbon dioxide, methane, or other
> greenhouse gasses is causing or will, in the foreseeable
> future, cause catastrophic heating of the Earth's
> atmosphere and disruption of the Earth's climate.
> Moreover, there is substantial scientific evidence that
> increases in atmospheric carbon dioxide produce many
> beneficial effects upon the natural plant and animal
> environments of the Earth.

Young strongly condemns the fact that the vast majority of
American scientists are not being heard by the media and

> is dismayed over the fact that the Global Warming fiasco
> has become politically popular and expedient to those
> liberal left politicians and power-brokers whose sole aim
> is to literally tax everything with a carbon footprint and
> give them control over all life, hidden within their PC
> guileful pretense to save the planet. They wish to save
> no one but themselves ... [the] entire IPCC process is
> but obfuscation by the secular and atheist Left. It has
> allowed the Left to conflate the vanity of secular opinion
> with scientific and/or moral truth.

In "Climate Deniers Are Giving Us Skeptics a Bad Name" (Singer 2012a), S. Fred Singer argues this extreme scientific taking of sides regarding climate change is dividing the universe of climate scientists into three groups.

> On the one side are the "warmistas," with fixed views about apocalyptic man-made global warming; at the other extreme are the "deniers." Somewhere in the middle are climate skeptics, who go along with the general conclusion of the warmistas but simply claim that the human contribution is not as large as indicated by climate models. But at the same time, they join with deniers in opposing drastic efforts to mitigate greenhouse gas emissions.

Singer goes into detail about how these three groups interpret the climate warming science published by UN IPCC reports, which conclude that most of the temperature increase in the last century is due to carbon dioxide emissions produced by the use of fossil fuels.

Being an expert reviewer of IPCC assessment reports, Singer provides credible insight into the scientific contributions being made to these reports. In his article, Singer discloses that he had an opportunity to review part of the Fifth Assessment Report (AR5), then due in 2013. From this review, he noted that the AR5 used essentially the same argument and evidence as the Fourth Assessment Report (AR4) published in 2007.

To avoid getting into the weeds as provided by Singer, he also provides for his readers a summary of three things wrong with the IPCC AR4 argument. First,

> It depends very much on detailed and somewhat arbitrary choices of model inputs [for example] the properties and effects of atmospheric aerosols, and their temporal and geographic distribution. It also makes

arbitrary assumptions about clouds and water vapor, which produce the most important greenhouse forcing. One might therefore say that the IPCC's evidence is nothing more than an exercise in curve-fitting.

The second and third problems according to Singer are these.

Can the IPCC fit other climate records of importance besides the reported global surface record? For example, can they fit northern and southern hemisphere temperatures using the *same* assumptions in their models about aerosols, clouds, and water vapor? Can they fit the atmospheric temperature record as obtained from satellites, and also from radiosondes carried in weather balloons? The IPCC report does not show such results, and one therefore suspects that their curve-fitting exercise may not work, except with the global surface record.

The third problem may be the most important and likely also the most contested one ... The IPCC conclusion ... depends crucially on the reported global surface warming between 1978 and 2000. As stated in their Summary for Policymakers (IPCC-AR4, vol 1, page 10): "Most of the observed increase in global average temperatures since the mid-20th century is very likely due to the observed increase in anthropogenic greenhouse gas concentrations."

Singer asks about issues raised by skeptics such as little to no earth warming between 1978 and 2000 and the data being collected from thousands of poorly distributed weather stations not representing a true global warming when in fact atmospheric temperature records between 1978 and 2000 from satellites and independently from radiosondes don't show warming. Neither does the ocean. And even the so-called proxy record from tree rings, ice cores, ocean sediments, coral, stalagmites, and others shows mostly no warming during the same period.

Singer exposes the false arguments being used by the deniers attempting to downplay alarmist arguments and

> concludes that both groups need to find a way to step back and learn to be responsible scientists by making their measurements, perfecting their theories, publishing their work, and hoping that in time the truth will out.

Subsequent to Singer's comments, one of its main conclusion of the final Fifth Assessment Report (AR5 2014) (IPCC 2014) reports was that,

> there was a 95 percent chance that human activity was affecting the climate. The AR4 report in 2007 had this number at 90 percent.

In "IPCC Now '95 percent Sure of Global Warming" (Folks 2013), Jeffrey Folks argues that the IPCC report does offer a detailed and to the non-specialist reader an impenetrable rationale for this change, but for Folks,

> These figures sound all too much like the old toothpaste commercial. The 'new and improved' product prevents 95 percent of all cavities—the old toothpaste prevented only 90 percent. Buy the 'new and improved'.

Folks argues that the AR5 reports suggests that "natural forces such as volcanic eruptions have overridden human activity so that warming has slowed or plateaued." Folks asks, "Just how much influence does human activity have, if any, when weighed against natural and cyclical forces?"

For Folks, this is not the question needing an answer; it is the expanding of powers of government that is precisely what is at stake in the battle over climate change. At the time, Folks speculated that the IPCC AR5 report would serve as fodder for the Obama administration's EPA and other agencies to intensify their war on coal and other carbon energy sources, which they did. Further, Folks speculates that climate science has the potential of robbing the earth's peoples of their liberty today.

Folks draws attention to the claim by the IPCC that its work is "never policy-prescriptive," (IPCC 2014), yet the AR5 report alludes to the role of energy and other industries in relation to carbon emissions and asserts that "it is likely or highly likely that man-made carbon emissions at current levels will lead to devastating consequences by the end of this century if not earlier." Folks then asks, "In what sense is this not 'policy-prescriptive'?" And while the panel claims to "reflect a range of views" (IPCC 2014), Folks could not identify a single major scientist who doubts the existence of man-made global warming.

Folks also noted that at the time of publishing the AR5 IPCC report, Mother Nature was also making it awkward for climate alarmists since it appears certain that the earth's surface climate is no longer warming citing the AR5 IPCC report, which documents that the northern hemisphere spring snow cover to now be greater than it was in 1990. All the IPCC can say about this trend is, "The earth is continuing to take up heat even when the surface is warming slowly."

Folks argues that it is no secret that the IPCC believes that the use of fossil fuels may be responsible for much of the warming that has taken place in the recent past. Further, it argues that at current levels of global carbon emissions, there is a chance of a 2 degree C increase in the earth's temperatures by 2040 or 2050.

> With no real way to prove the IPCC temperature projection, Folks speculates that it may also be natural cyclical forces that are more powerful than man-made influences and that these forces will balance out

whatever effect human behavior has on the earth's temperatures. Or it may be that temperatures will rise and that increase will be beneficial. But one thing is certain: more regulation will lead to higher energy costs, lower economic growth, and a reduced standard of living.

A cost-benefit analysis of global climate policy would see the squandering of trillions of dollars on restricting carbon emissions, which may or may not reduce future emissions. That would have a dramatic effect on global growth if allowed to remain invested in the private economy, the place history tells us pragmatic solutions will be found.

In their article "The IPCC is still wrong on climate change. Scientist prove it" (Dunn et al. 2018), John Dunn and Joseph Bast provide a critique of two more-recent reports on the current state of global warming that came to two completely different conclusions. The first was a special report (Coolearth 2018) released by the IPCC (IPCCSpecial 2018) that addresses the alleged impacts of "global warming of 1.5° C above pre-industrial levels and related global greenhouse gas emission pathways, in the context of strengthening the global response to the threat of climate change, sustainable development, and efforts to eradicate poverty."

The second report cited by Dunn and Bast, which coincided with the special report released by the IPCC, was released by the Nongovernmental International Panel on Climate Change (NIPCC 2018) and named the Summary for Policymakers of the fifth volume of its Climate Change Reconsidered (Heartland 2018) series.

According to Dunn and Bast, the two reports tell dramatically different stories about the causes and consequences of climate change. The IPCC report, referred to as SP15, claims,

> Human greenhouse gas emissions are causing an unprecedented warming of the planet's atmosphere, that it is too late to prevent a warming of 1.5° C above

pre-industrial levels, and that nothing less than a
dramatic reduction of the use of fossil fuels, possibly
even an outright ban enforced by the United Nations,
is needed to prevent a global catastrophe.

By contrast, Dunn and Bast note that the NIPCC report finds that
while climate change is occurring and a human impact on climate is
likely, there is no consensus on the size of that impact relative to natural
variability, the net benefits or costs of the impacts of

climate change, or whether future climate trends can
be predicted with sufficient confidence to guide public
policies today. Consequently, there is no scientific basis
for the recommendation that the use of fossil fuels
should be restricted.

Dunn and Bast cite the following key conclusions provided in
NIPCC's Summary for Policymakers.

Fossil fuels deliver affordable, plentiful, and reliable energy critical to human welfare. Wind and solar are not practical and reliable
substitutes.

Fossil fuels create a better environment for the ecosystem because
they require less surface area than renewable energy sources.

Sixteen of 25 identified impacts of fossil fuels are net positive, eight
uncertain. Only one is net negative. Areas of impact measured include
agriculture, air quality, extreme weather events, and human health.

Forcing a transition from fossil fuels to wind and solar
power would inflict tremendous economic hardship,
reducing world GDP by some 96 percent and plunging
the world back to economic conditions last seen in the
1820s and 1830s.

Dunn and Bast then address the question of how could two international teams of scientists, economists, and other experts arrive at opposite conclusions? They answer this question as follows.

> The IPCC is a political organization, not a scientific body. It was formed by the United Nations in 1988 for the purpose of establishing the need for a global solution to the alleged problem of anthropogenic climate change. Note that the mission of the IPCC was never to study the *causes* of climate change; were that the case, it might have devoted some of its billions of dollars in revenues over the years to examining solar cycles, changes in ocean currents, the sensitivity of climate to greenhouse gases, or the planet's carbon cycle. The IPCC has spent trivial sums on these issues, and the authors of and contributors to its voluminous reports have few or no credentials in these fields.

> However, the NIPCC is a scientific body composed of scholars from more than two dozen countries, first convened in 2003 by the great physicist S. Fred Singer and later chaired by another great physicist, Frederick Seitz. The NIPCC's only purpose is to fact-check the work of the IPCC. It receives no corporate or government funding and so has no hidden agenda or axes to grind. Most of its participants volunteer their time; a few receive token compensation for many hours of effort.

Further, Dunn and Bast argue,

> The NIPCC views the claim that human greenhouse gas emissions are causing climate change to be a hypothesis to be tested, not a preordained conclusion.

It asks whether the null hypothesis – that changes in climate are natural variability caused by a multitude of climate forcings and feedbacks – has been disproven. Its research reveals thousands of studies published in peer-reviewed science journals supporting the null hypothesis, meaning that the IPCC's mountains of data and expressions of "confidence" are irrelevant, meaningless, and ultimately wrong.

Finally, Dunn and Bast challenges their readers, "Given their provenances, which report do you think is more likely to be truthful?"

5

Computer Models, Chaos Theory, and Climate Change

Fanaticism consists in redoubling your effort when you have forgotten your aim. (George Santayana, *Life of Reason*, 1905, vol. 1, Introduction; US Spanish-born philosopher, 1863–1952)

IN 2016, Attorney General Loretta Lynch testified before the Senate that the Department of Justice was considering taking legal action against energy industries dubious of the dire role of carbon emissions to change the climate. Democratic attorney generals from numerous states are in hot pursuit of global warming heretics. Before more partisan lawyering and congressional testimony clouds the climate change concern, let's clear up what is known about this issue.

In "The Model Atmosphere and Global Warming" (Sadar 2016), Anthony J. Sadar cites the following quote from Prof. Paul Edwards: "Everything we know about the world's climate—past, present, and future—we know through models." According to Sadar, Edwards, a supporter of the consensus view of climate change, is quoted as saying in the introduction to his highly acclaimed *A Vast Machine: Computer Models, Climate Data, and the Politics of Global Warming* (Edwards 2010), "Without models, there are no data."

Sadar acknowledges that models have become integral to modern scientific practice. Edwards says that in many fields, "Computer models complement or even replace laboratory experiments; analysis and simulation models have become principal means of data collection, prediction, and decision making."

Sadar agrees that such is the contemporary world of science aided by the powerful tool of modern computers, and he reminds us that "the three basic components of the scientific method—observation, hypothesis, and testing—still hold, but in many cases, the testing portion has been abetted if not in some cases usurped by models." However, Sadar argues that many skeptics of the man-made disastrous global warming hypothesis attest that the evidence for a worldwide climate catastrophe is founded on the results of atmospheric models. Sadar then asks whether such results have been trusted enough to direct trillions of dollars in the years ahead to shift the energy sector and redistribute financial resources.

To answer this question, Sadar argues that

> atmospheric models have tremendous difficulty simulating key elements of the hydrologic cycle such as cloud cover and precipitation patterns. Water in all its phases—clouds, oceans, and condensed tiny droplets in clouds, ice, and snow—is the ultimate regulator of climate on earth with greenhouse gases such as carbon dioxide and methane playing a secondary role in climate control. However the research funding is on carbon pollution, and largely, only negative aspects of increased atmospheric carbon dioxide are headlined in the popular and scientific press.

Drawing from his thirty-five years of professional life as an air-pollution meteorologist, which he focuses on in his second article, "The Perspective of a Lifetime on Atmospheric Modeling" (Sadar 2012), Sadler explains his involvement with atmospheric modeling in one way or another. It's clear from his experience that the complexity of the earth's climate is vast. Forecasting the future of such climate in meaningful detail is incredibly challenging. Even really smart science sages cannot know the long-term state of the global climate other than in a wide range of temperature and precipitation levels, a range too wide to be of much practical value. This lack of knowledge has been confirmed over the past eighteen years as the modeled global average temperature trend has dramatically not matched reality.

Sadar reminds that when it comes to modeling Earth's distant future climate, perhaps the eminent atmospheric scientist, Reid Bryson, said it best: "Making a forecast is easy. Being right is the hard part."

In "Liberals and Mathematical Models" (Schmitt 2008a), Jerome J. Schmitt argues that climate alarmists and the liberal big media are placing unquestioning faith in mathematical models of man-made AGW, which support their predictions; they consider that settled science. They are fully assuming that the models completely and flawlessly account for all significant variables in one of the most complex systems ever studied, the earth's atmosphere.

For Schmitt, professional experience teaches climate scientists that modeling atmospheric phenomena is an uncertain business even in atmospheres much less complex than the earth's. To provide context to his argument, Schmitt makes reference to a far simpler artificial atmospheres or climates that are routinely created and modeled in the semiconductor industry, which uses sealed vessels within which chemical vapor deposition forms nanometer-scale thin solid films on silicon wafer surfaces to produce integrated circuits and many other semiconductor products.

In a supporting article, "Numerical Models, Integrated Circuits, and Global Warming" (Schmitt 2007a), Schmitt wrote about the many difficulties these simple atmospheric models encounter.

> Closed systems are also much easier to model as compared to systems open to the atmosphere (that should tell us something already). Computer models

are used to inform the engineering team as the design
the shape, temperature ramp, flow rates, etc, etc, (i.e.
the thermodynamics) of the new reactor.

Nonetheless for Schmitt, despite the facts that the chemical reactions are highly studied, that there exists extensive experience with similar reactors much of it recorded in the open literature, that the input gases and materials are of high and known purity, and that the process is controlled with incredible precision, the predictions of the models are often wrong and require that the reactor be adjusted empirically to produce the desired product with quality and reliability.

Despite the facts that these artificial climates are closed systems far simpler than the global climate is, that they have the advantage of the experimental method, and that they are subject to precise controls, they are frequently wrong. That should lend some humility to those who make grand predictions about the future of Earth's atmosphere.

Sometimes, so serious are the problems that it is not unheard of for an experimental reactor to be scrapped entirely in favor of starting from scratch in designing the process and equipment. Often, a design adjustment predicted to improve performance actually does the opposite. This does not mean that process models are useless; they undergird engineers' understanding of what is happening in the process and helps them make adjustments to fix the problem. But it means that they cannot be relied on by themselves to predict results. These new adjustments and related information are then used to improve the models for future use in a step-by-step process tested time and again against experimental reality.

Schmitt rightly notes that there is no consistent record of accurate predictions to date for the totally unproven and skepticism-worthy models of the global warming climatologists. However, climate alarmists and the liberal big media are convinced that there is a scientific consensus over an impending AGW crisis that requires taking immediate and extreme measures including increased carbon taxes as well as private-sector solutions such as carbon trading and selling carbon credits.

Due to this lack of success in predicting climate warming events by alarmists, the reader can reasonably assume that the cause can be traced to the construction, data-collection methods, and of course their interpretation of the stored data to produce their predetermined results.

In "The Granularity of Climate Models" (Thompson 2010), Bruce Thompson acknowledges that it has been hard to find the specific details of the construction of these climate models. Thompson asks his readers to keep in mind that at this time, computers are not electronic brains; they are just very fast electronic slide rules and adding machines.

Thompson notes that historic surface temperature data is maintained by a few government institutions; the more notable are the Hadley Climate Research Unit (CRU), an offshoot of the British Meteorological Office (the Met Office), and in the US at NASA's NOAA. The CRU was the home of the now-infamous Dr. Phil Jones made famous by the Climategate scandal.

The CRU developed the Hadley Centre Coupled Model, version 3 (HadCM3 2018), a coupled atmosphere-ocean general circulation model. This was one of the major models used in the IPCC's Third Assessment Report 2001 (IPPC 2001) and the Fourth Assessment Report 2007 (IPPC 2007), and it contributed to the Fifth Assessment Report 2014 (IPPC 2014). Thompson explains that the HadCM3 model has two components: the atmospheric model HadAM3 and the ocean model HadOM3 (which includes a sea ice model). Simulations use a 360-day calendar in which each month is 30 days.

Before any climate model can be built, it is necessary to break the earth's surface into three-dimensional numerical grids or cells. The atmospheric component (HadAM3) has nineteen levels with a horizontal resolution of 2.5 degrees of latitude by 3.75 degrees of longitude, which produces a global grid of 96 x 73 boxes (pressure, temperature, moisture). The vector grid (wind velocity) is offset by a half of a grid box; this is equivalent to a surface resolution of about 417 km x 278 km at the equator reducing to 295 km x 278 km at 45 degrees latitude. The oceanic component (HadOM3) has twenty levels with a horizontal resolution of 1.25 x 1.25 degrees. At this resolution, it is possible to represent important details in oceanic current structures.

Since the results are typically displayed as surface level only, to populate the two-dimensional grid box (longitude x latitude) with three variables (pressure, temperature, and moisture) for the 7,008 grid boxes, there needs to be enough readings to populate the grid boxes with surface variables.

According to Thompson, it is at this point that climate modelers are presented with their data integrity problem. Since they need data to populate 7,008 grid boxes, the maximum number of surface temperature stations in the databases was about 6,000, which has been culled to only about 1,200. This issue is addressed in the report "Corruption of ground-based temperature records used by the UN IPCC" (D'Aleo et al. 2010) by Joe D'Aleo and Anthony Watts.

Since the Hadley database is for land-based surface temperatures only, the data set of surface temperatures needed to populate its grid boxes could be reduced from 7,008 to 2,100 because the land surface of the earth is about 30 percent of the total surface. The existing 1,200 surface temperature stations are only a bit more than half of the land surface total required.

Thompson refers his readers to the website of eminent climatologist Dr. Roy Spencer (http://www.drroyspencer.com), who prepared the linked map entitled "ISH Surface Weather Stations Reporting 4+ times per day, 1986 thru 2009" (Spencer 2014) of the distribution of reporting stations. It shows the small number of reporting stations in the southern hemisphere. Thompson points out from studying this map that the continents of South America, Africa, Australia, and Antarctica are missing not to mention that 70 percent of the earth's surface covered by oceans has not been touched.

Thompson argues that even if the 1,200 stations were evenly redistributed across the earth's surface, the area of the average box would have to nearly double in light of the fact that the total number of grid boxes is an indication of the grid's granularity. The fewer boxes, the more granular the model becomes.

According to Thompson, to match the number of grid boxes to the number of stations, it is necessary to reduce the climate-reporting grid boxes to an eighty-by-fifty matrix creating 4,000 boxes—1,200 on land and 2,800 on the sea. The new grid box size would increase from 3.75 degrees longitude by 2.5 degrees latitude to 4.5 degrees longitude by 3.6 degrees latitude. According to Thompson, who cites Robert T. Merritt's estimate of the median radius of an Atlantic hurricane as 2.4 degrees of latitude, the new grid box is big enough to nearly swallow the median-sized Atlantic hurricane whole all the way to the outer-most radius of closed isobar (Wikipedia, radius). This would reduce an Atlantic hurricane to just one average temperature, pressure, and humidity value in the model. For Thompson, that would be absurd on its face.

Thompson then focuses on a single grid box identified in the Hadley climate model with the standard dimension of 3.75 degrees in longitude by 2.5 degrees in latitude.

For this, Thompson selects the grid box that encompasses Miami, which is at center long -81.875, lat 26.25, and corners at NW long -83.75, lat 27.50, SW long -83.75, lat 25.00, NE long -80.00, lat 27.50, SE long -80.00, lat 25.00.

Thompson cites the website NOAA's National Data Buoy Center (https://www.ndbc.noaa.gov), which provides a satellite view of the geographic area of interest and allows the user to find data in a given radius of the geographic center of the selected grid box. Using the center coordinate of long 81.875W, lat 26.25N and a distance of 100 NM (km), the website provides a list of stations safely in this one climate grid box. It is a circle of 200 NM (km or 230 statute miles) diameter. The site allows the user to move the cursor and get the longitude and latitude so the user can outline the grid box.

In selecting this one grid box and trying to distill all its texture information into one grain of data, Thompson concludes the global climate model is a "fool's errand". In the case of the Miami grid box, the first challenge is whether the grid box belongs to the land-based data set or the ocean data set with a bit more ocean than land. The added issue is the water surface topography, which includes the Gulf Stream, the Everglades, and the Gulf of Mexico. Each will have different water surface temperatures because sunlight can penetrate to much greater depths in the relatively still, clear, and deep Gulf of Mexico while its penetration in the Everglades is limited by the opacity of the water and its shallow depth. And the Gulf Stream is a conveyor of stored heat collected elsewhere in the tropics and released in Florida.

Finally, Thompson argues that because of all the gaps in existing data collection, climatologists are forced to perform what is called data homogenization; they need numbers to put in the empty grid boxes, so they average values from surrounding grid boxes that could be hundreds of miles away and fill them into the empty boxes.

Last, Thompson argues that currently, an accurate data set does not exist, so the climate model's input data and the resulting output can best be described in climate science terms as garbage in, garbage out.

Adding to the professional criticism of climate models as put forward by Thompson, Norman Rogers, in "The Global Warming Smoking Gun" (Rogers 2017a) argues that climate data populating the grid boxes of climate models include many approximations and assumptions that are set with adjustable parameters using tuning components that are not necessarily well grounded in atmospheric physics. Rogers argues the tuning components programmed into climate models are nondimensional coefficients or correction factors that are adjusted (tuned) so that the predictions of the equations fit real-world data with high fidelity. This is done to improve the models' accuracy for further predictions. Despite the fact that weather models may be tuned quite frequently with real-world data, they still cannot predict accurately beyond four or five days. For Rogers, climate models by contrast require many years or decades to pass before the predictions of multidecade climate models can be compared with empirical data for the purposes of such tuning.

The added complexity of climate modeling as compared to weather modeling is due to such factors as ocean currents, changes in vegetation, the global economy, and other factors that have no bearing on short-term weather that must be accounted for in climate models. The sets of mathematical relations are thus inherently more complex and require more equations to account for these extra parameters, and that leaves more room for error.

Because of the time interval needed to test the models' tuned components, Rogers notes that this testing is accomplished by running the models against the past readings and adjusting the parameters to make the model output agree with the known past climate. However, Rogers argues that the past climate is also not well known in many respects, so estimating is used, and different modelers have different past climate estimates. This tuning with the past may result in the model tuned to agree with the past but then failing to predict the future based on faulty assumptions or faulty data or both.

To provide evidence for his argument, Rogers cites the climate models used by the UN IPCC from its 2013 IPCC AR5 Report, in which they cannot be made to even agree with the past climate. As illustrated in this report (10.3.1.1.2), it plots the climate temperature observations against the averaged output of the various models used by the IPCC. There are two areas of serious disagreement illustrated by added annotation. From 1910 to 1940, the earth warmed strongly, but the models do not generate a match to that warming. The other area of disagreement is the period starting in 1998 when global warming stopped; that is called the hiatus or the pause. The models project global warming continuing, not stopping, in 1998.

For Rogers, the inability to explain the early twentieth-century warming and the real probability that the late twentieth-century warming may be forced by factors other than carbon dioxide constitute a smoking gun type of evidence that casts doubt on the predictions of global warming forced by carbon dioxide. Doubt concerning the viability of the climate models is further reinforced by the lack of warming during the last eighteen years, the hiatus.

In a number of articles, Brian C. Joondeph tackles the inability of climate modelers to provide trusted predictions of future weather conditions let alone the future climate. Joondeph acknowledges that weather and climate are not the same. In the publication "NASA—What's the Difference Between Weather and Climate?" (NASA 2005), NASA defines *weather* as representing atmospheric conditions over a short period while *climate* is measured over relatively long periods. Joondeph argues, however, that both use computer models in an attempt to predict the future.

As NOAA's Climate.gov website explains (Climate Models, NASA 2019), "Models help us to work through complicated problems and understand complex systems." Joondeph cites Dr. Jeff Masters's article "Hurricane Forecast Computer Models" (Masters 2019) to provide further insight into forecasting of weather and predicting future climate based on computer models.

Joondeph acknowledges that weather and climate are incredibly complex and as a result not easily predictable. One can predict tomorrow's weather by saying it will be the same as today's weather and be correct much of the time. Hurricanes in the next few days or climate in a century are not as easy to forecast. Because of this complexity, climate models have been developed to factor in measurements of temperatures on land, air, sea, ocean currents, wind patterns, geological activity, and a host of other variables. Although weather and climate are different disciplines, for Joondeph, they are common in that the resulting predictions of both rests on the collected data being fed into the model's equations. The modelers then need to interpret the results in such a way as to predict future events whether a hurricane over the next five days or the climate over the next five decades.

However, Joondeph argues that weather and climate cannot be predicted with accuracy at least given our current knowledge of climate.

The "Intergovernmental Panel on Climate Change's 2001 Assessment Report" (Laframboise 2010) agrees:

> The climate system is a coupled non-linear chaotic system, and therefore the long-term prediction of future climate states is not possible.

Joondeph explains the idea of chaos theory (WhatIs 2016) for complex systems such as weather, cloud patterns, financial markets, bird migrations, and so on. While appearing random, they actually follow a set of rules. Small changes in any of the numerous variables affecting such a system can change the outcome. The problem is that one can't measure every one of these variables or their seemingly inconsequential effects on the model. Think of a butterfly flapping its wings in Nepal and ultimately influencing a hurricane in the Caribbean. Heady stuff, but this explains why hurricane trackers can't know in advance where a current hurricane will make landfall; they can only guess. This is evident with dozens of hurricanes each year where each of its models produces a spaghetti-line prediction of where each hurricane will track as it progresses into the Gulf of Mexico, hits Florida, makes landfall as far north as Canada, or veers harmlessly out to sea.

So if models can't predict the track of a hurricane only days ahead, how does Gore and his minions of climate alarmists know what the temperature will be in 2050? Or the sea level? Or the mass of Antarctic ice, which as an amusing aside is increasing according to NASA in their posting "NASA Study: Mass Gains of Antarctic Ice Sheet Greater than Losses" (NASA 2015b). According to Joondeph, they can't. He supports his position by looking at prior climate predictions derived from models.

Anthony Watts posted an article "Defying Al Gore's predictions, bottom drops out of the US hurricane pattern over past decade" (Watts 2017) in which he cites climate scientist and soothsayer Gore making the following prediction in 2005.

> The science is extremely clear now, that warmer oceans make the average hurricane stronger, not only makes

the winds stronger, but dramatically increases the moisture from the oceans evaporating into the storm— thus magnifying its destructive power—makes the duration, as well as the intensity of the hurricane, stronger.

Joondeph notes Gore was not alone. *Scientific American* posted an article that asked, "Are Category 6 Hurricanes Coming Soon?" (Harper 2011). Aly Nielsen and Joseph Rossell in their article "Katrina Anniversary: Media's 10 most Outlandish Predictions Full of Hot Air" (Nielsen et al. 2015) recount examples of the liberal big media making such predictions as "No End In Sight For Big Hurricanes" and "Katrina Is The Beginning of What May Be A Long Stretch of Wild and Devastating Weather." Joondeph addresses these wild predictions by citing the history of hurricanes affecting the US from the same public information that would have been available to these activists making their wild prediction at the time.

First, Joondeph notes that the world has never seen (Fleshler 2018) a category 6 storm. The last category 5 storm to hit the US was Andrew in 1992. Joondeph cites the NOAA (AOML.NOAA 2018) compilation of hurricanes information going back 150 years. In the decade of the 2000s, the time of Gore's movie *An Inconvenient Truth* and his apocalyptic predictions, nineteen hurricanes impacted the continental US. All were categories 1 to 3 except for Charley in 2004, which was a category 4. There were four years in this decade with no hurricanes hitting the continental US.

By comparison, in the 1850s, a time before global warming was part of the popular lexicon, there was at least one hurricane each year, sixteen in all including a category 4 in 1856. In the 1880s, the continental US was hit by twenty-five hurricanes, one of them a category 4. The 1910s brought twenty-one hurricanes, three of which were category 4. In the 1940s, the US had twenty-three hurricanes with four reaching category 4.

Joondeph notes that recent decades show a decrease in total

number and powerful hurricanes compared to previous decades. This is contrary to what the computer models are predicting and what the media is warning about.

For Joondeph, it is ironic that a hurricane, representing a clear and present danger to millions of people, can have weather models that produce such unpredictable result but escape scrutiny. If Joondeph asserts that a named hurricane will head out to sea after models predict the hurricane will hit land, should he not be called a hurricane denier and threatened with prison as advocated in "US College Professor Demands Imprisonment for Climate-Change Deniers" (Owens 2014)? So if Joondeph or other skeptics are not being challenged when questioning projected tracks of next week's hurricane, why is there such visceral reaction to skeptics who question model climate predictions fifty years or a century from now?

6

Alarmists' Corruption of Climate Science

> If you tell a lie big enough and keep repeating it, people will eventually come to believe it. The lie can be maintained only for such time as the State can shield the people from the political, economic and military consequences of the lie. (Joseph Goebbels, http://www.HolocaustResearchProject.org)

IN A 2008 article, "Corrupted science revealed" (Schmitt 2008d), Jerome J. Schmitt states that "Outsiders familiar with the proper workings of science have long known that modern Climate Science is dysfunctional." At the time, a prominent insider, MIT meteorology professor Richard S. Lindzen, explains (Lindzen 2008a) how Gore and his minions used Stalinist tactics to subvert, suborn, and corrupt a whole branch of science and cites chapter and verse in his report "Climate Science: Is it currently designed to answer questions?" (Lindzen 2014). In detailing the corruption, Lindzen names names at the time of his report, one of which is as follows.

> For example, the primary spokesman for the American Meteorological Society in Washington is Anthony

> Socci who is neither an elected official of the AMS nor
> a contributor to climate science. Rather, he is a former
> staffer for Al Gore. (Lindzen 2014, 5)

Lindzen's report generally provides specific examples of how a radical cabal is forcing scientists to ignore or amend measurements that undermine the theory of AGW. According to Lindzen and other authors, scientists are literally forced to include sentences in their papers that indicate their support of AGW even if these sentences are non sequiturs or even if they conflict with the overall thrust of the paper. In this way, Gore's uneducated political commissars are able to deliver the consensus he and the alarmists advocate.

Along with Lindzen's arguments on how this could be possible, Schmitt provides his own personal experience as an undergraduate and graduate student in the hard sciences and later as a research collaborator with dozens of industrial scientists and university professors. Schmitt explains that today's scientists get to the top of their fields by extreme dedication to their specialty involving inordinate focus and concentration that cannot tolerate distractions. The best scientists are constantly at home at their lab bench with their instruments analyzing data, teaching a few promising students, and preparing publications. Most scientists interact intensively only with other specialists in allied fields.

Their dealings with one another are possible only by maintaining extreme standards of honesty, integrity, and open-mindedness to scholarly debate in search of the truth. The very qualities that make them good scientists and scholars thus leave them ill-equipped to deal with the raucous, underhanded, disrespectful, politically motivated radicals unleashed upon them by Gore and his fifth column for a hostile takeover of their scientific institutions.

Along with the majority of scientists, Schmitt naively believes that institutions such as the NAS can impose some quality control on an errant discipline as Schmitt outlined in another of his articles, "'Grantsmanship' Distorts Global Warming Science" (Schmitt 2008c).

Lindzen notes that this august body has been penetrated by ecoactivists by exploiting loopholes in its nominating procedures with the objective of corrupting its quality-control role.

Schmitt notes that for those who have wondered how liberal leftist cabals were able in the 60s and 70s to take over the universities' humanity departments, the National Endowment of the Arts, the National Endowment of the Humanities, and more important the Democratic Party and the liberal big media, Lindzen's report lays bare in his publication the template for their radicalization.

To provide more specifics in the evolution of climate corruption, in "Democrats' Real Global Warming Fraud Revealed" (Avery 2017), Dennis Avery describes pathbreaking findings by the European Organization for Nuclear Research (CERN, Wikipedia, CERN) reviving the sun as the controlling mechanism of climate and debunking the so-called global warming consensus. Perpetuators of the global warming myth had proposed that historical global average temperatures manifested a hockey-stick shape of sharply higher temperatures in the last decades of the twentieth century. Costly regulations were thus justified such as the Clean Power Plan (CPP, Wikipedia, CPP) promulgated by the US Environmental Protection Agency (EPA). The Clean Power Plan was unveiled by Obama on August 3, 2015; it aimed to reduce carbon dioxide emissions from electrical power plants by an additional 32 percent within twenty-five years from 2005 levels beyond those already achieved.

Trump's proposed 2018 budget, page 41 (Whitehouse 2018), eliminated the CPP and all related budget lines. Despite Trump's executive order, it is useful to review the manipulation of climate science used to justify the CPP and other climate-related policies in the first place.

The February 4, 2017 edition of the United Kingdom's *Daily Mail* carried David Rose's article "Exposed: How world leaders were duped into investing billions over manipulated global warming data" (Rose 2017). It exposed how Thomas Karl, director of the National Centers for Environmental Information (NCEI) at the NOAA and eight coauthors broke established NOAA and scientific protocols to rush the publication of a paper designed to influence debate over both the CPP and the December 12, 2015 Paris Agreement (Wikipedia, PA) on climate change. The pause-buster paper published June 2015 in the journal *Science* entitled "Possible Artifacts of Data Biases in the Recent Global Surface Warming Hiatus" (Karl et al. 2015) purported to refute a 2013 report from the UN IPCC (IPCC 2013), which found "a much smaller increasing trend over the past 15 years, 1998–2012 than over the past 30 to 60 years" (769, box 9.2) despite continued substantial increases of atmospheric carbon dioxide.

Dr. John Bates, another distinguished NOAA scientist, described to the *Daily Mail* two flawed data sets used in the pause-buster paper and violations of NOAA's and *Science* magazine's archival procedures. Bates also testified on February 5, 2017, before the US House of Representatives' Science Committee and commented in an article "Climate scientists and climate data" (Bates 2017) on the website of Judith Curry, former professor and chair of the School of Earth and Atmospheric Sciences at the Georgia Institute of Technology.

According to Bates, the first data set comprised ocean temperatures measured from ships by buckets and engine intake since the late 1800s and measurements from a system of buoys deployed only since 2000. The buoy data showed lower temperatures, and the ship data was widely considered less reliable because of selective sampling and the fact that ships warm surrounding water. However, Karl et al. "adjusted" the buoy data higher by an average 0.12 degrees centigrade and ignored measurements from satellites since 1979, considered highly reliable by scientists, that showed no warming after 1998. This resulting adjusted data set, the Extended Reconstructed Sea Surface Temperatures version 4 (ERSSTv4) thus created the appearance of continued warming.

The second NOAA data set, Global Historical Climatology Network (GHCN), is a time-series analysis of temperature readings from about 4,000 weather stations spread across the world's land mass. The GHCN data set was in a test phase, not ready for operational use.

Moreover, Bates discovered that the GHCN software included errors that made analyses unstable. Karl et al., however, concluded that past temperatures had been cooler than previously thought, again implying not a warming pause since 1998 but rather an increase.

Bates argues these misleading analyses were compounded by the failure of Karl and colleagues to properly archive data and analyses for other scientists to replicate—the gold standard of the scientific method. This is a violation of common standards set by both NOAA and *Science* magazine admitted by Karl to undermine scientific integrity. Therefore, the conclusions of the pause-buster paper are suspect, yet they were cited by the Obama administration in its implementation of the CPP.

This is not the only incident of data set manipulation by alarmist climate scientists. One of the original manufactured climate data sets is addressed by Dale Leuck in the article "Disingenuous Climate Science Debunked" (Leuck 2017) in which he details the leaked emails in November 2009 and November 2011 that were exposed in a scandal known as Climategate (Taylor 2011), which revealed unethical scientific behavior among a cohort of climate scientists attempting to influence international public policy. This was done through the UN's IPCC. The IPCC has produced a number of assessment reports since 1988 that are often cited as reflecting scientific consensus for the existence of global warming caused by human activities producing greenhouse gases and particularly carbon dioxide.

Leuck specifically cites the Third Assessment Report (IPCC-WGI 2001), which purports to show average global temperatures from AD 1000 with the average flat for over 900 years, trending upward around 1920, flattening around 1970, and then spiking higher in a so-called hockey-stick fashion. The hockey-stick graph included in the IPCC report was a theoretical reconstruction of global temperatures by Mann, Bradley, and Hughes (MBH 1998) (Mann et al. 1998) using statistical modeling based on inferences from a large sample of tree ring and ice core proxy data. The hockey-stick chart did not comport with traditional constructions of historical temperatures.

Leuck then cites McIntyre and McKitrick (MM 2003) (McIntyre et al. 2003), who debunked the hockey stick as derived from "collation errors, unjustified truncation or extrapolation of source data, obsolete data, incorrect principal component calculations, geographical mis-locations and other serious defects." After adjusting for such shortcomings, McIntyre and McKitrick applied MBH's (1998) methodology to the improved original-source database. Their analysis revealed higher northern hemisphere temperatures in the fifteenth century than in the twentieth consistent with traditional theory and contradicting the unusual results produced by MBH (1998).

Leuck then cites a 2006 scientific panel chaired by Dr. Edward Wegman of the National Academy of Sciences Committee on Applied and Theoretical Statistics. It also concluded that MBH's (1998) methodology was flawed. The Wegman (Wegman 2006) report submitted to the House of Representatives cited a National Research Council (NRC report SEM 2006) panel endorsing the results of MM 2003 and criticizing not only the statistical methodology of MBH (1998) but also its absence of collaboration with professional statisticians. Moreover, the Wegman panel criticized Mann and the IPCC for systematic unwillingness to share research materials, data, and results outside a small group of similar-minded analysts noting, "There was too much reliance on peer review which was not necessarily independent." The latter conclusion was based on a social network analysis of the seventy-five most frequently published authors in the area of climate reconstruction research to evaluate the true independence of research that reported results similar to MBH (1998).

Leuck further argues that not only was the MBH (1998) analysis debunked by MM (2003), the Wegman panel and the NRC, the MBH (1998), analysis itself also contradicted conclusions of the 1990 IPCC First Assessment Report (IPCC 1990) that illustrates roughly one and a half temperature cycles since about AD 900, when a warming phase began known as the Medieval Warm Period, which peaked around 1200. This was followed in the fourteenth century with the beginning of the roughly four-hundred-year-long Little Ice Age, during which advancing glaciers forced the abandonment of Viking settlements in Greenland and food shortages throughout Europe. The last warming cycle began in the late eighteenth or early nineteenth centuries that according to satellite and ground-based sensors continued until leveling off in 1998.

For Leuck, the IPCC Third Assessment Report was disingenuous in purging the Medieval Warm Period and thus creating the hockey stick as well as ignoring data suggesting that the late twentieth-century warm period may have not been particularly warm in a historical context.

Today, the main source of the current global temperature data sets is NOAA's NCEI, which published a temperature database from 7,280 worldwide meteorological stations; it is called the Global Historical Climatology Network (GHCN). In the article "The NOAA Database and Global Warming" (Long 2017), Edward R. Long explains each database consists of six files, three (maximum, minimum, and average) for unadjusted and three for adjusted values. The unadjusted values were formerly called raw data.

As noted by Long, the unadjusted values were as received from agencies/centers from around the world that collect the respective station data. For the United States, there are six collection centers known as NOAA's Regional Climate Centers (NOAA 2019). Supposedly, the unadjusted data are as-measured values, but the NOAA-NCEI advises on its GHCN web page, "It is entirely possible that the source of these data (generally National Meteorological Services) may have made adjustments to these data prior to their inclusion within the GHCNM."

According to Long, NOAA-NCEI makes its own adjustments, primarily a lowering of the temperature values of land stations in the earlier part of the temperature record. This adjusted data is used to determine global temperature trends and form the basis of the claim that global temperature is rising at a significant and concerning rate with the further claim that this warming trend is primarily attributed to human activities and specifically increased concentrations of carbon dioxide in the air from the burning of fossil fuels.

Long notes that numerous scientists, engineers, and others have challenged these adjustments and have asserted that the NOAA-NCEI's adjustments cause an artificial temperature increase. But this adjustment issue aside, Long argues there are at least three more-fundamental issues with the GHCN database. According to Long, they may be found by inspecting the GHCN unadjusted average-temperature database.

According to Long, a number of issues need to be addressed by NOAA-NCEI to record and provide reliable data for reporting to the public and to use for climate research. The first pertains to the accuracy of how these reporting stations are assigned to three location categories—urban, suburban, and rural—and whether each is at an airport. This categorization of each station is necessary since the resulting temperature reading will be biased by its geographic location, that is to say, a rural station at an airport with urban heat island readings being categorized as rural.

Long argues the second issue is more significant and pertains to stations involving airports. Long notes that NOAA-NCEI temperature stations show the yearly average temperature jumping approximately 2 degrees C about 1950. Some if not all of this discontinuity is due to the stations' reporting before 1950 and those reporting after 1950 being different stations at different locations. For example, stations pre-1950 being at airports not carried over to data post-1950 from all 7,220 stations. The net effect is a temperature trend interpreted as an indication of a rise in global temperature when instead the cause was the failure to take into account when stations at airports actually began to be at airports.

Long then notes the third issue, which he argues is perhaps far more significant than the first two. The durations stations report temperatures vary from as little as eight or ten years to nearly if not all years for a time frame, say 1900–2016. To exacerbate matters, some stations report data for only a few years in one portion of a time frame, and others at different locations report for only a few years in much different portions of the time frame. So for a particular time frame, station reporting pops in and out causing the averages to increase or decrease depending on the location.

To provide detailed analysis of the effects of these three reporting errors, Long has published a technical report (Long 2016) entitled "An Inspection of NOAA's Global Historical Climatology NetWork Verson 3 Unadjusted Data" (Long 2017).

In the article "NASA's Rubber Ruler: An Update" (Hoven 2018), Randall Hoven asserts, "The NASA/GISS temperature record is not actually a record of recorded temperatures. It is simply the most recent version of NASA's adjustments to older adjustments. It is not thermometer readings. It is models all the way down."

To support his argument, Hoven notes that in 2012, he wrote "Global Warming Melts Away" (Hoven 2012) on the status of global warming at the time. In it, he describes using the latest available NASA/GISS data to do his analysis, which was the version NASA had on its website as of April 30, 2012 (Land-Ocean Temperature Index). According to Hoven, this 2012 data for the period 1880–2011 showed a warming trend of 0.59 degrees C per century. Hoven then cites the warming trend using the December 30, 2017 data from NASA's website for the same period as being 0.66 C per century.

Because of this obvious change to the recorded temperature history, NASA believes the earth is getting warmer faster than it was five and a half years ago. For NASA, it is not because of actual recorded thermometer readings in those last five and a half years; it is getting warmer faster because NASA adjusted the data to show faster warming.

Hoven further explains that a researcher can download temperature anomalies "1880–present" by going to NASA's website. However, according to Hoven, NASA adjusts this data every month. Most disturbing for Hoven is that researchers will not find any older versions; NASA makes available only its most recent version and does not explain how it adjusts the data.

Hoven explains that he has actual proof of this data manipulation by NASA based on an actual data set he had downloaded to a spreadsheet in 2012 to support the updated article he wrote in 2012 (cited above). In his article, Hoven provides graphs that show all adjustments to data by NASA after April 2012 for the period 1880–2011. Hoven makes it clear that the trend is of the adjustments to temperatures, not actual temperatures, and he argues that NASA tends to adjust older temperatures down and recent temperatures up to accelerate the overall warming trend from 0.59 to 0.66 C per century just since 2012.

Hoven notes that looking only at the most recent century of those same data (1911–2011), the adjustment trend is even starker: from 0.71 to 0.87 C per century. Again, the only difference is when the data were downloaded from the NASA website.

In his "rubber ruler" (Hoven 2012) article in 2012, Hoven wrote of NASA's changing the temperature record going back to 1880 every month. In just one month in 2012, August to September, 60 percent of NASA's temperature records changed.

In "Delingpole: 'Nearly All' Recent Global Warming is Fabricated, Sturfu Finds" (Delingpole 2017), James Delingpole argues that this has the effect of exaggerating the warming trend. A peer-reviewed study by Joe D'Aleo, Cato Institute climate scientist Craig Idso, and statistician James Wallace—two scientists and a veteran statistician—looked at the global average temperature data sets, which are used by climate alarmists to argue the recent years have been the hottest ever and the warming of the last 120 years has been dramatic and unprecedented. In addressing these alarmists' claims, the authors found these readings are "totally inconsistent with published and credible U.S. and other temperature data."

In "NASA GISS caught changing past data again—violates Data Quality Act" (Watts 2012), Anthony Watts outlines how NASA is in violation of the Data Quality Act in the way it is managing this critical climate data, which is being used to study the uncertainties of climate change.

For Hoven, researchers cannot validate a climate model using temperature observations if those observations were themselves being adjusted using models. Real science means employing the scientific method—using physical measurements to test a hypothesis. Hoven argues that NASA is reversing that method. It apparently uses the global warming hypothesis to adjust physical measurements. That is not science. It is the opposite of science.

In another example of manufactured data to support the climate alarmists claims of man-made climate change, in "Warmist Cargo Cult Science Returns" (Birdnow 2011), Timothy Birdnow cites Michael Mann, Pennsylvania State University climate Svengali, who figured prominently in the Climategate email scandals at England's University of East Anglia and was one of the creative authors of the famed hockey-stick graph that was so influential on the 2007 IPCC Climate Change report. Mann coauthored "Climate related sea-level variations over the past two millennia" (Kemp et al. 2010), which claims to show a massive acceleration in sea-level rise in North Carolina that coincides with the industrial era.

According to Birdnow, this study claims to have reconstructed 2,000 years of sea levels. Further, Birdnow interprets the study to actually extrapolate from a study of shallow salt marshes with a historical reconstruction going back 300 years and based on the prevalence of *foraminifera* fossils to reconstruct past sea levels. These reside in the very shallow, sandy pools and die in deeper waters, so theoretically, researchers could see where sea levels were in the past. This study then used tide gauge data to calibrate the data. By observing agreement between direct observations and this proxy reconstruction, the authors estimated the rate of rise and extrapolate into the past.

Sarcastically, Birdnow states that the authors were exhaustive in their methodology in choosing a whopping two points (Sand Point and Tump Point) to study the fossils and calibrating from data from two other points (Wilmington and Hampton Roads). Their conclusion from this limited empirical research is that "sea level rise has accelerated, and this 'correlates' to the industrial era." Birdnow cites from this study's abstract.

> Sea level was stable from at least BC 100 until AD 950.
> Sea level then increased for 400 y at a rate of 0.6 mm/y,
> followed by a further period of stable, or slightly falling,
> sea level that persisted until the late 19th century. Since
> then, sea level has risen at an average rate of 2.1 mm/y,
> representing the steepest century-scale increase of the
> past two millennia. This rate was initiated between AD
> 1865 and 1892.

Birdnow further cites the posting "Louisiana Coastal Land Loss" (Davis-Wheeler 2000) in which Clare Davis-Wheeler outlines that Louisiana has lost considerable coastal marshland as a result of human intervention. Flood-control dams and levees prevent swollen rivers from picking up silt, and dredging to keep the coastal waterways open moves the silt from its natural place leading to erosion of the coastal shallows. According to this paper, "The main forms of human disturbance are the river-control structures such as dams and levees, the dredging of canals, and draining and filling."

According to Bardnow, a large part of the sediment gathered by existing marshes is accumulated during seasonal flooding. Flood overtopping and overbank sedimentation, both vital to the survival of existing marshes, were dramatically reduced as large areas ceased to be flooded. River water also helped to reduce marsh salinity and provide nutrients, and its loss has resulted in the breakup and dispersal of large amounts of nutrient-starved marshlands. Without the extra silt brought from floodwaters, the shallows are subject to erosion and breakup. This would clearly warp the fossil record but would also warp the tide gauge record as well; that is, the sea would appear to be rising when in fact the land is sinking, being washed away.

In "Further Problems with Kemp and Mann"(Eschenbach 2011), Willis Eschenbach provides a map (Watts 1990) of the North Carolina sites from 1733 juxtaposed with a satellite photo from 1990. The reader will notice the radical difference between the two.

Birdnow notes that erosion is a problem here, yet the authors of the Kemp and Mann paper fail to give it any credence. Of course, the fossil record will show sea-level rise!

Birdnow notes that as to the correlation between the industrial era and this increase in rise rate, the rate increase appears to begin around 1880, well before the rise in industrial emissions. It would not be before an increase in land-use change that would contribute to erosion. Birdnow also emphasizes that older tide gauge data is likely to be poor resulting in the older data being suspect. According to Willis Eschenbach (Eschenbach 2011), it is not even consistent with itself. Along with a number of uncertainties with the Kemp and Mann study sea rise conclusions, Eschenbach notesthat as is not uncommon with sea level records, nearby tide gauges give very different changes in sea level. In this case, the Wilmington rise is 2.0 mm per year, while the Hampton Roads rise is more than twice that, 4.5 mm per year. In addition, the much shorter satellite records show only half a mm per year average rise for the last twenty years.

Birdnow claims that this sea-level study by Kemp and Mann has averaged two records and contradicted all other studies that show a far lower sea-level rise. For Birdnow, this is not science; it is advocacy in costume. There was a predetermined outcome, the sites of study were chosen with that outcome in mind, and the authors issued a big, glossy press release before the publication of the paper to make a splash with the salivating dogs (Peracchio 2011) of the liberal big media. The authors knew this would be analyzed to death, but they wanted it out before the public first. Likely, the public would hear that, yes, sea levels are rising faster and would get little of the rebuttal.

Despite the deliberate manufacture of fake science, people such as Mann continue their climatological malpractice and are even accorded respect. Such shoddy workmanship in any other field would put the principal into another line of work.

But not in climatology; being a charlatan hack seems to be de rigueur.

7

Climategate Exposes
Alarmist Conspirators

ON NOVEMBER 17, 2009, the email hacking scandal that has become known as Climategate exposed to the world documents, source code, data, and emails in a folder purportedly hacked from Britain's University of East Anglia (UEA) Climate Research Unit first uploaded to a Russian FTP server in the wee hours of November 17, 2009, and announced that evening as a comment at Air Vent (Id 2009), which revealed a widespread pattern of scientific misconduct among the very climate researchers on whose science the AGW theory and all consequent policy is based.

In "Climategate: One Year and Sixty House Seats Later" (Sheppard 2010d), Marc Sheppard comments that

> this dossier of hacked incriminating e-mails that exposed the specter of governance and wealth redistribution both national and international in the 1990's based largely, if not solely, on pseudo-scientific hocus-pocus should have ended the AGW debate abruptly and evermore. Remarkably it didn't.

Sheppard argues that at the time the Cancun's "last chance to save the planet" climate talks were just around the corner and the liberal big media were working overtime to explain away previous failures as any-thing other than the product of bad policy toward unproven hazards that they indeed were.

Sheppard cites the *Washington Post* article "Coverage of climate summit called short on science" (Reuters 2010), which focused on an Oxford University's Reuters Institute study as to who attended this event and how countries covered the 2009 UN Summit. But the *Washington Post* article's emphasis was somewhat different and clearly divulged in its headline. Yet for Sheppard, what truly boggles the mind is the assessment.

> Much coverage from Copenhagen instead focused on hacked e-mails from a British university that some skeptics took as evidence of efforts by scientists to ignore dissenting views. The scientists involved have since been cleared of wrongdoing.

For Sheppard, it was Climategate that exposed the conspirators' arrogant mockery of the peer-review process atop a widespread com-plicity in and acceptance of hiding, manipulating, inventing, and oth-erwise misrepresenting data in a clear effort to exaggerate the existence, causation, precedence, and threat of global warming. The fact that many of the conspirators were editors, lead authors, and contributors to the UN IPCC and WMO reports on which international climate policy is made puts all such reports and policies in question.

As the news-media-led understanding of the Climategate folder's incriminating contents took wider purchase, formal investigations were launched as the evidence of climate fraud appeared both devastating and incontrovertible ("The Evidence of Criminal Fraud," Sheppard 2009a).

By August 2010, the final reports were in from all three formal investigations into Britain's Climategate at the University of East Anglia Climate Research Unit: the House of Commons Science and Technology Committee (HOC 2010), the Oxburgh Science Appraisal

Panel (UEA 2010), and the Independent Climate Change E-mails Review under Sir Muir Russell (ICCER 2010).

Sheppard notes that all three examinations took place in the country of physical jurisdiction, Great Britain, and not one of them disappointed those scientists anticipating a whitewash of the exposed Climategate offenses. Simply stated, all parties were cleared of all wrongdoing other than perhaps sloppy journaling and sophomoric note-passing, and all suspensions were lifted. As Andrew Montford summarized in his report *The Climategate Inquiries* (Montford 2010),

> There can be little doubt that none of [the inquiries] have performed their work in a way that is likely to restore confidence in the work of CRU. None has managed to be objective and comprehensive. None has shown a serious concern for the truth. The best of them—the House of Commons inquiry—was cursory and appeared to exonerate the scientists with little evidence to justify such a conclusion. The Oxburgh and Russell inquiries were worse.

Sheppard then cites the investigation undertaken by a Pennsylvania State University Inquiry Committee into the specific actions of the institution's employee, Dr. Michael Mann. Based in the US, the Penn State inquiry offered perhaps the best hope of impartiality. After all, not only was a faculty member implicated at the deepest levels of the misconduct ("Understanding Climategate's Hidden Decline," Sheppard 2009b); he was also implicated in the attempt to destroy evidence.

According to Sheppard, it was Mann's fellow Penn professors who were tasked with investigating him, and there was scant shock in the scientific community when the Inquiry Committee's unanimous determination came out: "Dr. Michael E. Mann did not engage in, nor did he participate in, directly or indirectly, any actions that seriously deviated from accepted practices within the academic community for proposing, conducting, or reporting research, or other scholarly activities."

For Sheppard, there is little question that the initial silence and ultimate dismissal by the liberal big media was and remains a factor in Climategate's surprisingly marginal effect by liberal left policy makers. But the societal and financial impacts of proposed policies shaped by the misinformation in question are nothing short of astounding.

As Ross McKitrick stated in "Understanding the Climategate Inquiries" (McKitrick 2010),

> The world still awaits a proper inquiry into Climategate: one that is not stacked with global warming advocates, and one that is prepared to cross-examine evidence, interview critics as well as supporters of the CRU and other IPCC players, and follow the evidence where it clearly leads.

Sheppard believes it was the absence of any authoritative investigation particularly in the US that likely provided the greatest cover of all to climate alarmists around the world. As well, it is *Washington Post* articles such as one written by Bracken Hendricks, "Don't believe in global warming? That's not very conservative" (Hendricks 2010) that made the outrageous claim, "The best science available suggests that without taking action to fundamentally change how we produce and use energy, we could see temperatures rise 9 to 11 degrees Fahrenheit over much of the United States by 2090."

Sheppard asks his reader to apply a little common sense to news articles such as this one when considering Hendricks's claim. To meet his temperature projection, the global temperature would need to warm every decade for eight decades by about the same amount the IPCC claims the globe warmed in all of the previous century, reverse the current global temperature warming pause since 1998, and do so soon. It is important to note that Hendricks made his claim in 2010, and there has been no dramatic change in global temperatures.

Further, Sheppard argues that Climategate's initial revelations of corruption proved to be just the beginning. In the months that followed, allegations of similar misconduct among alarm-leaning climate scientists throughout the globe arose almost daily. Their affiliations were as momentous as those of Jones, Mann, and Briffa et al. and included the NOAA and the NASA/GISS ("Climategate: CRU Was But the Tip of the Iceberg," Sheppard 2010a), and ultimately the IPCC ("IPCC: International Pack of Climate Crooks," Sheppard 2010b) itself.

In "Just Sign on the Dotted Line" (Wright 2010b), Dexter Wright exposes the Jones gang in its effort to misinform nations by hiding their climate change facts and overstating the degree of consensus knowing that secrecy is essential for propaganda to be effective and ensure that the checks continue to be signed. Their propaganda allowed the group to convince the big liberal media of their global warming theory's legitimacy by the warnings supposedly signed by large numbers of the world's climate scientists.

To drive home this point, Wright cites several leaked Climategate emails that reveal backstage manipulations to produce a propaganda tool, the Statement of European Climate Scientists on Actions to Protect Global Climate (Climatecost 2011), which at the time was intended to be unveiled at the Kyoto Climate Conference. Members of the Jones gang from East Anglia University organized efforts to get just about anyone to sign this statement to push up the numbers. In an email dated October 9, 1997, Dr. Joseph Alcamo admonishes other members of the Jones gang to forget credentials and just get signatures.

> I am very strongly in favor of as wide and rapid a distribution as possible for endorsements. I think the only thing that counts is numbers. The media is going to say "1,000 scientists signed" or "1,500 signed." No one is going to check if it is 600 with PhDs versus 2,000 without. They will mention the prominent ones, but that is a different story.

According to Wright, Alcamo clearly has no respect for the media implying that they were either lazy or stupid. Operating under this premise, Alcamo said, "Conclusion—Forget the screening, forget asking them about their last publication (most will ignore you.) Get those names!"

Wright then notes that simultaneously, the folks at Greenpeace were also working to get signatures on a document of its own to manipulate the media. Their formula is tried and true: don't read the fine print—just sign. To showcase this subterfuge, Greenpeace was organizing a media event ahead of the Kyoto meeting to display the document signed by concerned scientists. The Jones gang wanted to make sure that maximum media manipulation was accomplished by coordinating media events as is detailed in the same email.

> If Greenpeace is having an event the week before, we should have it a week before them so that they and other NGOs can further spread the word about the Statement. On the other hand, it wouldn't be so bad to release the Statement in the same week, but on a different day. The media might enjoy hearing the message from two very different directions.

According to Wright, one of the Jones gang was not quite with the thinking of the group when Prof. Richard Tol, in an email dated November 12, 1997, pointed out the dirty little secret that there was not a consensus among scientists. Wright said,

> I am always worried about this sort of things. Even if you have 1000 signatures, and appear to have a strong backup, how many of those asked did not sign? ...Why was so much energy put into a propaganda campaign for the media to see that there was a 'consensus' among the scientific community?

According to Wright, the answer dates back to 1992, when the Jones gang was caught by surprise right before the Earth Summit in

Rio de Janeiro. At that time, a group of notable and respected scientists began circulating a document known as the Heidelberg Appeal (APC 1992) for signatures. By the end of the 1992 summit, 425 scientists and other intellectual leaders had signed the appeal. This document stated that the science of climate change was uncertain and that the theory of carbon dioxide–induced global warming was an unproven theory. The document appealed to policy makers to avoid making policy based on uncertain science. The document explicitly stated the following: "We do, however, forewarn the authorities in charge of our planet's destiny against decisions which are supported by pseudoscientific arguments or false and non-relevant data."

Wright comments that the original Heidelberg Appeal was presented at the Rio conference, but it was largely ignored by the liberal big media and a pseudoscientific community that was more interested in seeking grant funding than the truth. To date, more than four thousand scientists and intellectuals from 106 countries, including seventy-two Nobel Prize winners, have signed it. Wright argues the Jones gang knew that this would likely happen again before the 1997 Kyoto Climate Conference. If they were right, they were hopeful that they could deliver a counterdocument to lend credence to their cause and steal the spotlight.

Wright cites another document urging caution that was circulated among reputable scientists in the wake of the Kyoto Climate Conference. This document is known as the Leipzig Declaration on Global Climate Change (Leipzig 2005). The document expressly states the following.

> As the debate unfolds, it has become increasingly clear that—contrary to the conventional wisdom—there does not exist today a general scientific consensus about the importance of greenhouse warming from rising levels of carbon dioxide. In fact, most climate specialists now agree that actual observations from both weather satellites and balloon-borne radiosondes (i.e. weather

balloons) show no current warming whatsoever—in direct contradiction to computer model results.

Among the signatories of this declaration are scientists from NASA, the Max Planck Institute, a former president of the National Academy of Sciences, and many members of the American Meteorological Society. These people are not lightweights in the field of science.

In addition to these two powerful and well-considered public statements calling for restraint, there is also the Oregon Petition (Oregon 2007). To date, over 31,000 American scientists have signed this document. The petition explicitly states,

> There is no convincing scientific evidence that human release of Carbon dioxide, Methane or other greenhouse gasses is causing, or will in the foreseeable future, cause catastrophic heating of the Earth's atmosphere and disruption of the Earth's climate.

Unlike the exposed Climategate emails from the Jones gang, these statements of caution are in the public domain and have been for years.

The more damning actions of the Jones gang were their attacks on climate research scientists when the research did not support the Jones gang's fake climate warming hypothesis. As an example, in "Climategate: How to Hide the Sun" (Wright 2010a), Dexter Wright notes that the Climategate crowd successfully worked to obscure the connection between solar activity and climate, which the leaked CRU emails revealed as well.

Wright provides some background. In 2003, two Harvard-Smithsonian professors, Willie Soon and Sallie Baliunas, published the peer-reviewed paper Global Warming (Soon et al. 2003) in the scientific journal *Climate Research*. It identified solar activity as a major influence on Earth's climate. This paper also concluded that the twentieth century was not the warmest nor was it the century with the most extreme weather over the past thousand years. These two scientists reviewed more than two hundred sources of data and specifically climate variations observed to coincide with solar variations.

One of the more notable correlations cited in this paper is the well-documented coincidence of the Little Ice Age and a solar quiet period from the years 1300 to 1900. Soon and Baliunas asserted that the lack of solar activity resulted in cooler temperatures across the globe. The evidence they compiled also indicated that as the sun became more active, global temperatures began to rise and the Little Ice Age ended.

Wright comments, "In the past, the issue of the solar connection has always fallen down on one question; what is it about sunspots that cause a change in the climate?" Soon and Baliunas identified this physical connection as solar wind, which varies on an eleven-year cycle similar to sunspots. The solar wind is made up of high-energy particulate radiation, and when it is strong enough, it has a visible effect upon the atmosphere in the form of auroral displays in the polar regions (the Northern Lights). Some instances of solar wind were so powerful that the aurora could be seen even in lower latitudes.

Even with such convincing evidence, the Soon and Baliunas paper became the target of a great deal of criticism from the Jones gang at the CRU at East Anglia University. The hacked Climategate emails from Jones and his collaborators show an orchestrated effort to discredit the work of these two scholars.

Wright notes the discussion of solar influences being brought up in an email from Daly dated August 9, 1996. Daly uncovered an eleven-year signal in the temperature data set from the island of Tasmania. He found this signal by using a mathematical signal analysis formula known as a Fourier transform. It is clear from the tone of his email that he knows this is not welcome news, but he states the following concerning the temperature data set compiled by the Jones gang: "I tried the same run [Fourier transform] on the CRU global temperature set. Even though CRU must be highly smoothed by the time all the averages are worked out, the 11-year pulse is still there, albeit about half the size of Sydneys." The identified eleven-year climate cycle ("How the Sun's 11-Year Solar

Cycle Works," Chow 2011) corresponds exactly with the one observed on the sun. This fact was kept secret by the Jones gang.

Wright observes that the Soon and Baliunas references were the same data the Jones gang had reviewed and suppressed. The data in question is known as proxy data. Proxy data is data compiled from tree rings, sediments, and ice cores as well as other indirectly measured estimates of temperature. Correlating the timeline of these proxy data was identified as problematic by Wigley, another member of the Jones gang, in an email dated August 12, 1996. In his effort to correlate the data, Wigley concludes that the solar signal is strong enough to convince him that the sun is a major factor in climate change.

> Causes. Here, ice cores are more valuable (CO_2, CH4 and volcanic aerosol changes). But the main external candidate is solar, and more work is required to improve the "paleo" solar forcing record and to understand how the climate system responds both globally and regionally to solar forcing.

For Wright, what is significant about this paragraph is that it identifies the main cause of climate change as solar forcing, not carbon dioxide. This fact was also kept secret.

Remarkably, this was exactly what Soon and Baliunas published in their *Climate Research* paper. The solar correlation became a lightning rod. More than a dozen emails from the Jones gang discuss how to discredit Soon and Baliunas. Ultimately, the gang decided to compile a new paper to counter the conclusion made by Soon and Baliunas as detailed in an email from Dr. Scott Rutherford dated March 12, 2003. Rutherford does not go head to head with the data presented in the *Climate Research* paper, but he seemingly wishes to cook other data to counter the honest work of Soon and Baliunas.

> First, I'd be willing to handle the data and the plotting/ mapping. Second, regarding Mike's suggestions, if we

use different reference periods for the reconstructions and the models we need to be extremely careful about the differences. Not having seen what this will look like, I suggest that we start with the same instrumental reference period for both (1xxx xxxx xxxx). If you are willing to send me your series please send the raw (i.e. unfiltered) series. That way I can treat them all the same. We can then decide how we want to display the results.

Rutherford suggests that Soon and Baliunas should be dealt with severely: "There is nothing we can do about them aside from continuing to publish quality work in quality journals (or calling in a Mafia hit)."

Wright comments that it seems clear that the Jones gang felt threatened by the *Climate Research* paper. By all appearances, they saw the threat as significant enough to consider the scientific equivalent of evidence-tampering to hide the sun.

A 2010 preliminary report, "Climategate: Leaked E-mails Inspired Data Analyses Show Claimed Warming Greatly Exaggerated and NOAA not CRU is Ground Zero" (D'Aleo 2010) by Certified Consulting Meteorologist Joseph D'Aleo indicted a broader network of conspirators and challenged the very mechanism by which global temperatures are measured, published, and historically ranked. D'Aleo uncovered compelling evidence that the US government's principal climate centers have also been manipulating worldwide temperature data to fraudulently advance the global warming political agenda.

In a radio interview on KUSI-TV to discuss the "Climategate—American Style" scandal, D'Aleo and computer expert E. Michael Smith cited the alleged perpetrators as the NOAA and the NASA/GISS.

These two researchers accused NOAA of strategically deleting cherry-picked, cooler-reporting weather observation stations from the temperature data it

provides the world through its National Climatic Data
Center (NCDC). D'Aleo explained to show host and
Weather Channel founder John Coleman that while
the Hadley Center in the U.K. has been the subject
of recent Climategate scrutiny, "[w]e think NOAA is
complicit, if not the real ground zero for the issue.

According to D'Aleo, their primary accomplices are the scientists
at GISS, who put the altered data through an even more biased regimen
of alterations including intentionally replacing the dropped NOAA
readings with those of stations in much warmer locales.

As Sheppard explains in "Climategate: CRU Was But the Tip of
the Iceberg" (Sheppard 2010a), the ultimate effects of these statistical
transgressions on the reports that influence climate alarm and subsequently world energy policy are nothing short of staggering. According
to Sheppard, though satellite temperature measurements have been
available since 1978, most global temperature analyses still rely on data
captured from land-based thermometers scattered more or less about
the planet. NOAA receives and disseminates that data but not before
performing some adjustments to it.

Sheppard explains that NOAA argues it makes these adjustments
to the cherry-picked recording stations' data supposedly to eliminate
flagrant outliers, to adjust for time-of-day heat variance, and to homog-
enize stations with their neighbors to compensate for discontinuities.
According to NOAA, this homogenizing is accomplished by essentially
adjusting each to jibe closely with the mean of its five closest neighbors.
But given the limited number of stations and the likely disregard for
the latitude, elevation, or urban heat islands of such neighbors, it's no
surprise that such homogenizing seems to always result in warmer
readings.

NOAA's bogus data is accepted as green gospel as are its equally
bogus hysterical claims such as this one from the 2006 annual State of
the Climate in 2006 (WMO 2006): "Globally averaged mean annual

air temperature in 2005 slightly exceeded the previous record heat of 1998, making 2005 the warmest year on record."

As D'Aleo points out in the preliminary report, the 2010 NOAA proclamation that June 2009 was the second-warmest June in 130 years will go down in the history books despite multiple satellite assessments ranking it as the fifteenth coldest in thirty-one years. This same misinformation is provided by the National Weather Service when it makes its frequent announcements that a certain month or year was the hottest ever or that five of the warmest years on record occurred in the last decade; it's basing such hyperbole entirely on NOAA's warm-biased data.

But by far the most significant impact of this data fraud is that it ultimately bubbles up to the pages of the climate alarmists' bible, the UN IPCC Assessment Reports.

As argued by Sheppard in "Climategate's Phil Jones Confesses to Climate Fraud" (Sheppard 2010c), all of these schemes are to discredit skeptic climate warming scientists. Their objective is to control energy consumption while redistributing wealth, which relies on the same like-minded Jones gang of scientific experts to justify their price tags that are likely measured in the trillions.

Sheppard concludes that Climategate has firmed many of these expert scientists at NASA, NOAA (Sheppard 2010), and even the IPCC itself (Sheppard 2010), who have breached the standards of both science and ethics to advance their global warming political agenda.

Be that as it may, the level of deception perpetrated by Jones and his coconspirators at the CRU will likely remain the yardstick by which climate fraud is measured for years to come.

8

The Liberal Left's CO$_2$ Climate Alarmist Agenda

THE REPORTING by liberal politics and their complicit liberal big media over the past thirty years is unmistakably focused almost exclusively on the science of CO$_2$ forcing as the major cause of global warming to the exclusion of any other atmospheric warming agent including the sun if that is possible. This extreme bias could happen only with a complicit liberal big media and of course the agenda-driven, money-hungry liberal politicians.

In "Obama on the 'urgency' of combating 'global warming'" (Monckton 2008), Viscount Monckton of Brenchley cites a video (Obama 2008) that was shown at a costly, two-day global warming jamboree at the Beverly Hills Hotel hosted by then Governor Schwarzenegger of California in November 2008. From this video, Monckton cites Obama saying,

> Few challenges facing America and the world are more urgent than combating climate change. The science is beyond dispute and the facts are clear. Sea levels are rising. Coastlines are shrinking. We've seen record drought, spreading famine, and storms that are growing stronger with each passing hurricane season.

> Climate change and our dependence on foreign oil, if
> left unaddressed, will continue to weaken our economy
> and threaten our national security.

Obama said he would introduce "a federal cap and trade system to reduce America's emissions of carbon dioxide to their 1990 levels by 2020 and reduce them an additional 80 percent by 2050." And his administration would "invest" $15 billion a year in solar power, wind power, biofuels, nuclear power, and clean coal to "save the planet" by creating 5 million new "green jobs."

Monckton provides science-based details on the questionable assertions made by Obama during this video and throughout his eight-year administration.

Obama and the climate alarmists are not correct to say, "The science is not in dispute." In 2008, across all disciplines, some 32,000 scientists approached by the Oregon Institute of Science and Medicine in 2007/8 signed The Oregon Petition (Oregon 2007), which in effect states that global warming is not a global crisis and that humankind has very little influence over the climate. A survey of climatologists and scientists in related fields by Van Storch (2005/6) established that a considerable proportion of respondents did not believe the alarmist notions disseminated by Gore or the UN climate change panel.

The Obama administration and the climate alarmists were erroneous when they suggested that global warming would weaken the US economy and threaten national security. Ironically, Obama then began implementing through an executive order and a Supreme Court ruling the view that carbon dioxide was a pollutant and measures that would eventually cut 80 to 90 percent of US carbon emissions, which would have led to the closing down of 80 to 90 percent of the nation's industries, fatally weaken the US economy, and reduce it to Third World status. It is disingenuous of Obama and the climate alarmists to suggest that millions of new green jobs would appear out of nowhere to replace the scores of millions of jobs that the full implementation of his proposed measures would destroy.

As noted by Charles Battig in his 2015 article "Global Warming and Government Work" (Battig 2015), the comments made by these political leaders have been imbued with visions of pending climate catastrophes and have elevated the climate vagaries of Mother Nature to astronomical importance.

Against the 2015 global backdrop of multiple tribal warfare, televised images of beheadings, civil unrest, mass murders, and death threats acted out against the Western concepts of civilization, liberal politicians preached an even greater threat. Then Secretary of State John Kerry (Romm 2014) concluded after surveying the global landscape, "Climate change [is] perhaps the world's most fearsome weapon of mass destruction."

Battig cites Obama's 2015 State of the Union address (Obama 2015) wherein he states,

> The best scientists in the world are all telling us that our activities are changing the climate, and if we do not act forcefully, we'll continue to see rising oceans, longer, hotter heat waves, dangerous droughts and floods, and massive disruptions that can trigger greater migration, conflict, and hunger around the globe ... The Pentagon says that climate change poses immediate risks to our national security. We should act like it.

He described (Laymann 2015) rising temperatures as "the greatest threat to future Americans."

Battig then cites another world leader, Pope Francis, who despite all these worldwide security concerns announced that he has singled out climate change to be the topic of a papal encyclical (Lamb 2014) in 2015 "to make all people aware of the state of our climate and the tragedy of social exclusion."

For Battig, with so many political and religious reputations at stake, these leaders are looking anxiously to find some hard numbers from the science community to support their exaggerated climate concerns.

However, it is because of Mother Nature's nonpartisan pragmatism that she is not cooperating with these wild assertions either in the past or going forward into the future. What a tough spot to be in for those climate scientists whose task it is to divine a correlation or temperature trend to support alarmists' claims on which hang the fate of the world and future generations.

To provide context for his argument, Battig makes reference to the US government's stewards of global temperature at the NOAA when it announced on January 16, 2015 (NASA 2015a), "The year 2014 ranks as Earth's warmest since 1880, according to two separate analyses by NASA and National Oceanic and Atmospheric Administration (NOAA) scientists ... NOAA provides decision makers with timely and trusted science-based information about our changing world," said Richard Spinrad, NOAA chief scientist.

Further, Battig cites NASA confirming these findings.

> The 10 warmest years in the instrumental record, with the exception of 1998, have now occurred since 2000. This trend continues a long-term warming of the planet, according to an analysis of surface temperature measurements by scientists at NASA's Goddard Institute of Space Studies (GISS) in New York.

With this degree of confidence in a major government-run temperature collection agency, Battig sarcastically asks, "Who could now be so callous as to harbor any thoughts regarding the merits of 'government work'?"

According to Battig, NASA, after garnering headlines around the world documenting a febrile planet, again made headlines when GISS's director Gavin Schmidt (Rose 2015) "admitted NASA thinks the likelihood that 2014 was the warmest year since 1880 is just 38 per cent." Battig argues this discrepancy has to do with making accurate measurements of something rather tiny as if someone were trying to measure the thickness of a sheet of paper with a yardstick. These scientists belatedly acknowledged that it would be hard to trust the accuracy of such measurements ... something to do with esoteric confidence bands and the like ...

For Battig, all is not lost. This agency has now put a number (real scientists like to do that) on the quality of government work. Henceforth, getting something correct to 38 percent of certainty is the new benchmark for "good enough for government" work.

In "O'Leary casts an bleary eye on 'Eco-nutters'" (Osorio 2007), Ivan Osorio cites an interview by the *Wall Street Journal* with Michael O'Leary, CEO of the low-fare Irish airline Ryanair. During the interview, O'Leary offered an interesting array of comments on the airline industry and the regulatory environment, but he saved up his most scathing attacks for the new climate change taxes with which Britain was hitting the airlines. Among other things, O'Leary laid bare the whole global warming business for the fraud that it is.

> The problem with all this environmental claptrap ... It's a convenient excuse for politicians to just start taxing people. Some of these guilt-laden, middle-class liberals think it's somehow good: "Oh, that's my contribution to the environment." It's not. You're just being robbed—it's just highway [bleeping] robbery.

Fast forward to 2019 and another example of these tax grabs being imposed on an oil-rich country—the Canadian liberal government and the liberal chatter class have been passing legislation to tax Canadians with their disguised Carbon Tax program with big liberal media running a near-endless parade of stories to hammer home this highway robbery scheme better known as global warming taxes.

It is not enough for the climate alarmists, however, to promote their climate change nonsense through the government and liberal big media; its propaganda is been spread by environmental NGOs into universities and K-12 schools. However, the scariest influencers in this doomsday tactic are the unelected, unaccountable judges in the legal system who have been striking down and creating new social and environment laws.

As an example of this legal creativeness, in "'Future Generations' Sue the USA over Global Warming" (Clancy 2018) T. R. Clancy cites a 2018 stay of a "landmark" (SCOTUS 2018) Supreme Court trial ordered by Chief Justice John Roberts in the federal lawsuit against the US and various executive agencies filed on behalf of twenty-one children "and future generations" (Siegel 2018). According to Clancy, *Juliana v. United States* (Aiken 2015) alleges violations of the children's fundamental right to "a climate system (SCOTUS 2015) capable of sustaining human life." Acting as the guardian for the minor plaintiffs is James Hansen (Harris et al. 2017), the climate change equivalent of Patient Zero in the pandemic of terror over global warming.

To provide background on this case, Clancy cites the organization that initiated this particular example of lawfare as Our Children's Trust (OCT 2019), which claims to provide a voice for youth in its mission to "protect earth's atmosphere" preferably through "legally binding, science-based" climate recovery policies. *Juliana* illustrates the organization's idea of legally binding controls over American energy policy: the plaintiffs are requesting a court order directing the US government to "cease [its] permitting, authorizing, and subsidizing of fossil fuels and, instead, move to swiftly phase out CO_2 emissions."

Clancy quoted from "Young People Are Suing the Trump Administration over Climate Change. She's Their Lawyer" (Gonchar 2018) in the *New York Times*. Justice Department attorney Jeffrey H. Wood is calling the lawsuit an "unconstitutional attempt to use a single Oregon court to control the entire nation's energy and climate policy." In its petition to stay the case, the US argued that the lawsuit is "an attempt to redirect federal environmental and energy policies through the courts rather than through the political process, by asserting a new and unsupported fundamental due process right to certain climate conditions."

Clancy noted that even Obama's Justice Department fought this case (*Juliana* was filed in 2015) warning in its earlier stay request against a court empowered "to make and enforce national policy concerning energy production and consumption, transportation, science and technology, commerce, and any other social or economic activity that contributes to carbon dioxide ... emissions."

Clancy argues that if his readers are skeptical that a lone Oregon US district court judge could really end up with that kind of power, he makes reference to a lone California US district judge holding US border enforcement in her hand (Dinan 2018) while other black-robed solo acts have decreed how things will be with DACA, the Trump travel ban, transgenders in the military, and sanctuary cities (Sessions 2018). Clancy further reminds us that the Supreme Court's discovery of a previously unnoticed fundamental right in the Constitution to so-called marriages between members of the same sex came down to how Anthony Kennedy (SCOTUS 2014) felt about the idea.

At the time this book was being published, there had been no decision by the Supreme Count on this matter; however, it is being cited here to demonstrate how easy it is to alter social and economic policies in a democracy without the intervention of elected officials.

For Leo Goldstein ("President Trump: Time to Abolish Climate Alarmism in America," Goldstein 2017), the notion of the world burning up at some unspecified future date has been challenged by Trump's refusal to bow to climate alarmism and has caused sincere alarm among the climate alarmists and exposed their agenda as much bigger than it seems; it has consumed hundreds of billions of dollars annually and is demanding trillions more. For Goldstein, the US finally has a president who is willing and able to abolish climate alarmism in America. The only questions are when and how.

Goldstein argues there is no middle ground. Attempts to appease climate alarmists with statements such as "We agree with your concerns, but let us decide on the pace of the actions" have not only failed to check the climate agenda but have strengthened it as well. In "Renounce Climate Alarmism" (Goldstein 2017), Goldstein argues that climate alarmism renouncement must include declaring independence from the IPCC, which has an explicit mandate to pervert science, and he suggests a review of activities that promoted its self-professed authority in the US.

For Goldstein, indecision and compromise confuse friend and foe alike. Trump cannot receive support from the public without making a firm stand. Goldstein predicts that when Trump renounces this climate warming scam, things will change. Many smart people will take a harder look at the science, the purported scientists, and the whole process. Witnesses will come forward when the wall of silence is broken and the victims and witnesses feel no fear of retaliation.

However, Goldstein recognizes that climate alarmism will not simply fade away. Something receiving hundreds of billions of dollars annually cannot fade away with very powerful political forces having tied their destinies to climate alarmism. In the US, these forces include Big Green, the miseducational complex, and even the Democratic Party. Abroad, most of the European political establishment is on the hook.

Although not renouncing climate alarmism and jumping the chasm by outright rejecting the pseudoscience that concludes carbon dioxide as the only forcing atmospheric element causing climate warming, Trump has withdrawn the US from the Paris Agreement (Wikipedia, PA) on June 1, 2017, which is "an agreement within the United Nations Framework Convention on Climate Change (Wikipedia, UNFCCC) dealing with greenhouse gas emissions mitigation" (Wikipedia, CCM). It was signed originally by Obama with 194 other countries.

In "Trump Steps on the Paris Agreement, Stands for Sovereignty" (Ludwig 2017), E. Jeffrey Ludwig notes that Trump's announcement was one of the most defining moments in US history. The president spent considerable time listing the many ways the Paris Agreement is bad for the US economy, a burden on taxpayers, and insignificant as far as protecting the US and world environment. He announced his intention to be environmentally friendly but not to the point of shipping US jobs overseas, putting a lock and key on American energy resources, or curtailing US prosperity and quality of life.

Ludwig argues that the economic facts would have been enough to justify the US's withdrawal from the Paris Agreement. For Ludwig, the defining moment was when Trump made the following statement.

> There are serious legal and constitutional issues as well. Foreign leaders in Europe, Asia, and across the world, should not have more to say with respect to the U.S. economy than our own citizens and their elected representatives, thus, our withdrawal from the agreement represents a *reassertion of America's sovereignty (emphasis in original)*. Our constitution is unique among all nations of the world. And it is my highest obligation and greatest honor to protect it. And I will ... It would once have been unthinkable that an international agreement could prevent the United States from conducting its own domestic economic affairs, but this is the new reality we face if we do not leave the agreement or if we do not negotiate a far better deal.

Ludwig takes the position that Trump was announcing to the world that the US was departing from the trajectory of the US toward globalization. "America First" and the sovereignty of its nation-state are defined by its borders and as a self-governing entity answerable to its citizens and its national values and government system.

According to Ludwig, Trump is heeding George Washington's warning about the republic entangling itself with international alliances because Washington understood that the US is a unique country founded on unique republican principles, the rejection of all titles of nobility, and rights for all guaranteed by a constitution. Further, the US's rich soil and active commercial life provided great opportunities for all unlike the rigid feudalism that still characterized much of Europe in his time. He did not want the US to become compromised by other peoples who were less free and less prosperous than were the citizens of the US.

Sadly, the Washington-Lodge-Borah vision of the US as a unique land of opportunity and as a sovereign nation unique in many positive ways has been fading from view since the end of World War II. After the war, the UN (1945) was created. Then, the International Monetary Fund and the World Bank were created as international lending institutions (to which the US was and is the main contributor) to support the economic advancement of developing nations in the Third World.

NATO came into existence in 1947 to defend Europe against Soviet expansion. The US joined SEATO in 1954 to protect southeast Asia. Step by step, the network of multilateral memberships grew. Economic global agreements and networks also became the norm. The US became signers of the General Agreement on Tariffs and Trade (GATT) begun in 1947, which by 1994 had 128 signatories (WTO 1994) and which is now managed through a framework known as the World Trade Organization (WTO). Under GATT, procedures are in place for nation-states to negotiate disputes if they believe that the rules of GATT are not being followed or to challenge other perceived unfair trade practices.

During the years of President Bill Clinton, the US became one of the three members of the North American Free Trade Agreement (NAFTA), which Trump has been working to renegotiate and ratify.

This incredibly extensive military and economic network of treaties and agreements is the background against which we can see and understand the radical nature of Trump's repudiation of the Paris Agreement. He is thus speaking against not merely membership in the Paris Agreement and speaking for US sovereignty; he is also throwing down the gauntlet to its entire strategy of world relations during the post–World War II period and forthrightly bucking a seventy-two-year trend toward multilateralism, a seventy-two-year trend of diluting American sovereignty. He is saying no to a furtherance of the many financial and legal compromises the US made when entering such extensive networks. Ludwig cites that Trump with great clarity closed his announcement by saying, "In other words, the Paris framework is just a starting point. As bad as it is. Not an end point."

To provide evidence that Trump's instincts are serving him well, James Taylor, in "Election Slaughter for Climate Activism" (Taylor 2018), argues that as a result of the 2018 midterm elections,

> voters throughout the country inflicted a bloodbath on climate activism and climate activist political candidates. According to Taylor, voters rejected the two highest-profile climate activist ballot initiatives, severely punished Republicans who joined the congressional Climate Solutions Caucus, and sent home the Democratic climate activist most heavily supported by billionaire environmentalist Tom Steyer.

At the state level, Taylor cites that Washington State voters soundly rejected a ballot initiative that would have taxed carbon dioxide emissions. It was the second consecutive election in which Washington voters rejected a carbon dioxide tax. Even in deep-blue Washington, the proposed tax lost by double digits in 2016 and again in 2018.

Taylor then cites Arizona voters, who inflicted a humiliating defeat on a ballot initiative that would have required 50 percent of the state's electricity to come from renewable sources. Voters rejected the initiative by a whopping 70 percent to 30 percent. The defeat was particularly crushing to Steyer, who wrote the ballot initiative and spent approximately $20 million supporting it.

Taylor did acknowledge that the only climate alarmists' silver lining came in Nevada, where 59 percent of voters approved the same 50 percent renewable-power mandate that Arizonans rejected. However, to amend the Nevada Constitution, voters must approve an amendment in consecutive elections, which means they'll have to approve the proposed amendment again in 2020. Although Nevadans rendered first-round approval, it was made possible only because Steyer spent

$6 million backing the measure and there was very limited organized opposition. In light of Arizona's massive rejection of the same proposal, this Steyer initiative faces an uphill climb in 2020.

Taylor noted that voters also demolished the notion that climate activism can be a winning message for Republican politicians. The congressional Climate Solutions Caucus, which claims to seek and support "economically viable options to reduce climate risk," lost half its Republican membership. That means that twenty-two of the forty-three Republicans—more than half the caucus's Republican membership—disappeared after the elections. This included sending caucus cofounder and Republican leader Carlos Curbelo home in defeat despite the power of incumbency, a huge spending advantage, and the benefit of facing a novice political opponent.

Republicans were not the only elected officials to feel the wrath of the voters. Taylor noted that Democratic climate activists also suffered crushing defeats. The biggest was the defeat of Florida gubernatorial candidate Andrew Gillum. Global warming activists have long argued that global warming is a winning political issue for Democrats and Republicans alike in Florida. Gillum made global warming a central issue in his campaign, and Steyer donated $7 million to Gillum's campaign.

Alaskan voters also delivered a stinging rebuke to another Democratic climate activist. Global warming activists argue that Alaska, like Florida, is particularly fertile political ground for politicians to emphasize global warming. Nevertheless, Alaskans elected Republican Mike Dunleavy governor instead of Democratic climate activist Mark Begich, who had taken an even more aggressive position on global warming than incumbent Governor Bill Walker.

Taylor argues that the reason climate alarmists failed on election day isn't because Americans don't care about being good stewards of the planet; it's because the more people have a chance to examine the evidence related to the causes and consequences of climate change, the less concerned they are about global warming.

For Taylor, there is no obvious evidence that humans are on the verge of suffering from catastrophic, man-made climate change. Voters know it, and that's why they sent a clear message to Democrats and Republicans in the midterms: stop putting the interests of liberal left environmental radicals before those of working families, and if you do, do so at your own peril.

9

The Scam of Sustainable Dumb Energy

In "Sustainable Nonsense" (Folks 2011), Jeffrey Folks argues that "large-scale central planning, even of the most ruthless sort, has always been the left's preferred mode of operation." The former Obama administration's planning focused on achieving what the liberal left calls sustainability. Like so many core values of the liberal left, sustainable policymaking is a dominant role for government.

As the great political writer George Orwell understood, the devaluation of human life is always accompanied and enabled by the corruption of language. As Orwell put it in "Politics and the English Language" (Orwell 1946), "The decline of a language must ultimately have political and economic causes."

Folks argues that when applied to energy, sustainability has almost nothing to do with the ability of certain fuels to meet the nation's needs over an extended time. With the advent of hydraulic fracking, America possesses enough reserves of natural gas to meet its energy needs for at least a century. The US reserves of coal are even greater, but neither of these fuels is deemed clean enough by environmental groups to be sustainable.

For Folks, the environmental lobby and the former Obama administration had the odd notion that sustainability could be brought about by restricting consumption alone. Driving a compact hybrid with a reported fuel economy of fifty mpg city and forty-eight mpg highway may seem to some a better, more sustainable option.

However, for the liberal left, sustainable conjures up a source of power that can be deployed indefinitely and that produces zero emissions, not even carbon dioxide. Folks notes that as currently used, sustainability applies only to solar, wind, and geothermal sources of energy. He disagrees and argues that none of these sources are sustainable—they are not even viable because of their high price and inherent unreliability and the geographical footprint required to install vast solar and wind farms. Ironically, the least sustainable sources of power are those championed by the environmental movement.

Folks speculates that for the liberal left, the ultimate solution is to legislate the removal of human beings from part or all of the earth's surface. More than a few environmentalist leaders, including Obama's former national science and technology advisor John Holdren, advocated the reduction of human population to what they consider a sustainable level.

Folks argues that the word *sustainable* is meant to hide the fact that fossil fuels will be needed and used for a long time into the future and that eventually new sources of energy, unlikely to be solar or wind in their current forms, will supply the domestic and industrial needs of countries. A rational energy policy would junk the notion of sustainability altogether and proceed instead from the reality that civilization requires cheap and reliable sources of energy to fuel its ever-expanding needs. While fossil fuels can meet most of those requirements for the next century, conventional energy sources may well be supplanted in time by undreamed-of alternatives and currently the stuff of science fiction.

Supporting Folks's arguments, Stephen Moore and Kathleen Hartnett White in *Fueling Freedom: Exposing the Mad War on Energy* (Moore et al. 2016) expose the myths promulgated by renewables cheerleaders and their extensive misinformation campaign on clean-energy resources, the supposed low-environmental impact alternatives to fossil fuels, by arguing against what the authors propose would be disastrous—an energy production shift that would have serious lifestyle and geopolitical consequences for free-market countries.

Historically, Moore and Hartnett remind us that until coal was harnessed on a massive scale, humans were dependent on energy from plants, wood, animals, and human muscle as well as wind and water. The dramatic shift from diffuse and variable flows of energy—wind and water—to massive stores of hydrocarbon minerals was a turning point for human progress. Energy became transportable, controllable, affordable, plentiful, reliable, and versatile.

Moore and Hartnett school their readers on how the Industrial Revolution fueled by carbon energy use broke through decades of static human existence and brought significant historic and upward trends for the average person including a tripling of life expectancy and a ten- to thirtyfold increase in per-capita real income. Coal and petroleum transformed into energy for mechanical power were the most important energy conversions in industrial civilization. With coal-powered machines, humanity was suddenly liberated from the physical limitations of muscle and beasts of burden. When electricity became available, heat, power, countless household appliances, industrial motors, and electronics were developed that generated a second energy revolution.

Moore and Hartnett argue that it was the invention of the internal combustion engine that brought liberty, mobility, and choice that enabled sustained productivity and economic growth. Additionally, it revolutionized the science and practice of metallurgy and dramatically transformed textile production. Previously expensive and tedious to produce, clothing became more affordable and warmer; winter clothing became available. Today, 60 percent of global fibers come from fossil fuels. In addition, fossil fuels played and continue to play an important role in reducing food-supply loss by refrigeration, packaging, and containers.

Fossil fuels have also dramatically benefitted agriculture. Moore and Hartnett argue that US food production has tripled but is using a third of the land, labor, and cost of pre-fossil fuel agriculture. In the past, over 50 percent of the US population was involved in agriculture, and food was scarce and expensive. Today, only 3 percent of the country's population produces our plentiful food supply.

To provide further perspective, Moore and Hartnett note that non-fossil fuel energy sources—wood, wind, solar, hydro, geothermal, biomass, and nuclear—make up only 15 percent of the world's total primary energy supply and provide significantly lower energy yields and potential.

For comparison purposes, the power density (power per unit of volume) of natural gas–fired electric generation is almost 2,000 times greater than that of wind-generated electricity. Using ethanol produced from corn to power a vehicle's internal combustion system creates a net energy loss when the energy used for planting, fertilizing, harvesting, distilling, and transporting is factored in. Further, the diversion of 40 percent of the US corn crop to ethanol, a less efficient fuel than gasoline, has raised corn prices and prompted more farmers to grow corn instead of other vital crops while increasing tractor and other farm vehicle usage and chemicals that increase pollution.

In "Big Green Energy Machines" (Ausubel 2004), Jesse H. Ausubel says that the production of wind energy requires the average wind system to use 460 metric tons of steel and 870 cubic meters of concrete per megawatt. By contrast, a natural gas–combined cycle plant uses about three metric tons of steel and twenty-seven cubic meters of concrete.

Concerning the driving narrative behind the sustainable energy hoax and its objective to reduce carbon dioxide pollution, Moore and Hartnett assert that even with a 200 percent increase in GDP, air quality actually began falling in the 1960s, nearly a decade before the establishment of agencies such as the US EPA. These improvements have come from emission reductions and controls made by private business rather than by EPA mandates. As noted by Moore and Harnett, between 1980 and 2010, airborne sulfur dioxide declined by 89 percent, carbon monoxide by 82 percent, nitrogen dioxide by 52 percent, ozone by 27 percent, particulate matter by 27 percent, and mercury by 65 percent. Over the past few decades, tailpipe emissions declined by more than 90 percent while miles traveled increased by 180 percent.

Despite these obvious quality of life and environmental advancements, the poorly informed public is bombarded daily with climate alarmists' propaganda that renewable energy is the way of the future and that fossil fuels will be the cause of the earth's destruction. The renewable energy advocates solution, according to Spencer P. Morrison in "Wind and Solar Energy Are Dead Ends" (Morrison 2017), is to boldly claim that the world could be powered by renewable energy as early as 2030 ("Graph of the Day: Why 'experts' get it wrong on wind and solar," Parkenson 2014)—with enough government subsidies that is. And of course, the liberal big media, (Renewable energy smashes global records in 2015 report shows, Neslen 2016) play their part by hyping up the virtues of solar and wind energy as the solution to climate change.

Morrison acknowledges that renewable energy advocates are quite right; in terms of generational capacity, wind and solar have grown by leaps and bounds in the last three decades—wind by 24.3 percent per year since 1990 and solar by 46.2 percent per year since 1990.

According to the US Energy Information Administration (EIA) projects (Doman 2016), world energy consumption will increase by 48 percent by 2040. With this increase in energy demand, Morrison argues that despite their rapid growth, wind and solar energy have not and will not make a dent in global energy consumption. In fact, after thirty years of beefy government subsidies, renewable energy sources account for only approximately 3.7 percent of all energy consumption (EIA 2019).

Again using IEA data, Morrison notes that between 2013 and 2014, global energy demand grew by 2,000 terawatt-hours. Citing "Wind Energy Meets Just 0.46 percent Of Global Energy Demand—Despite Hundreds Of Billions In Investment" (NEE 2017b), Morrison argues that to meet this increasing demand, there would be a need to build 350,000 new two-megawatt wind turbines, which would entirely blanket the British Isles. For context, Morrison cites the above article (Ridley 2017) in which the author argues that this increase in wind turbines is 50 percent of those that have been built globally since the year 2000. Wind power is not the future; there is simply not enough extractable energy, and we are currently getting very close to that physical limit.

Morrison also cites "Here's how much of the world would need to be covered in solar panels to power Earth" (Harrington 2015) and argues that the state of solar energy is only slightly more promising. Harrington suggests that to meet the growing energy needs using solar, it would be necessary to cover an equatorial region the size of Spain with solar panels to generate enough electricity to meet global demand by 2030. This is an enormous amount of land that could otherwise be used for agriculture or left pristine; it also underestimates the size of the ecological footprint since only 20 percent of humankind's energy consumption takes the form of electricity. Should the world switch to electric vehicles, the area needed would be five times as large.

To complicate the viability of these renewable energy sources, Morrison argues that it would not be possible due to resource limitations. For example, each 1.8-square-meter solar panel requires 20 grams of silver ("You can't have solar without silver," DiLallo 2014). Since there are 1 million square meters in a square kilometer, 11.1 tons of silver are needed per square kilometer of solar panels. Spain is 506,000 square kilometers. Covering this much space with solar panels would require over 5.6 million tons of silver. As it turns out, that is over seven times as much silver as is estimated to exist in Earth's crust ("A 100 percent Solar-Powered Future Is Impossible—Requires 7.2 Times More Silver Than Currently Exists," NEE 2017a). This same logic applies to dozens of other minerals needed for solar panels and wind turbines; it is simply not feasible on a large scale because solar panels are resource-hungry. Morrison reminds his readers that modern electronics require many of the same minerals as do solar panels and wind turbines.

In short, renewables are not making a difference. Because of this, Morrison argues it would be far more cost-effective and reasonable to take the massive subsidies paid to the renewable energy industry and simply invest in more energy-efficient technology and its research.

However, the defenders of wind and solar claim that government subsidies are a minor help to get a new industry going and counter arguments by climate skeptics with the false claim that fossil fuels receive huge subsidies ("The Big Green Lie—Fossil fuels are massively subsidized," Helmer 2015). In reality, the fossil fuel industry pays huge taxes to the government ("Oil Industry Taxes: A Cash Cow for Government," Hodge 2010).

Norman Rogers ("Wind and Solar Energy: Good for Nothing," Rogers 2018c) provides an understanding of the subsidies being provided to wind and solar. The explicit subsidies include such things as a 30 percent construction subsidy for solar and a 2.3¢ cent per kilowatt hour subsidy for wind. Both technologies benefit from tax equity financing, a scheme based on special tax breaks and gaming the corporate income tax of a highly taxed corporate partner.

Yet according to the US Energy Information Administration ("Levelized Cost of Electricity and Levelized Avoided Cost of Electricity Methodology Supplement," EIA 2013), a better way to measure the wind and solar subsidies is to look at the benefits and losses to the economy. Rogers notes that a net loss to the economy implies a subsidy. Once it is recognized that a subsidy is present, the next step is to figure out who is paying for it. Invariably, it is either the taxpayer or the consumer of electricity.

Rogers notes that without subsidies and in locations with good wind or sunshine, the cost of producing wind or solar electricity is about seven cents per kilowatt-hour ("Cost of wind, Solar and Natural Gas Electricity at the Plant Fence," Dumbenergy 2018). By coincidence, the cost is almost equal for the two technologies. Since these technologies don't require fuel, most of the cost is the amortization of the capital investment and each unit's maintenance, which calculates to seven cents per kilowatt-hour and is competitive with coal or nuclear but more expensive than natural gas.

But unlike conventional generating plants, wind and solar produce erratic electricity that comes and goes depending on wind and sunshine. Rogers explains that because of its unreliability, wind and solar plants cannot displace conventional plants because the conventional plants have to stay in place as backup plants to supply electricity when the erratic wind or solar is not producing electricity. Although it is often claimed that wind or solar is replacing conventional generation, it only reduces the operating duty cycles of the conventional plants. The backup plants are usually natural gas plants because natural gas plants are agile and able to follow the rapid ups and downs of wind or solar better than are other types.

For Rogers, the economic benefit of wind or solar is in the fuel savings in the backup plants when backup electricity is displaced by wind or solar electricity. The cost of fuel for a natural gas plant is about 2¢ per kilowatt-hour. The difference, the 7¢ cost of generating wind or solar electricity less the 2¢ benefit for fuel saved, is a 5¢ per kilowatt hour subsidy for wind or solar. Rogers estimates that about 3¢ of the 5¢ difference is probably accounted for by the explicit government subsidies; the other 2¢ is paid for by the consumer.

Rogers explains that for residential rooftop solar, the numbers are even more gruesome. In "Dumb Energy—A Critique of Wind and Solar Energy" (Rogers 2018b), he provides some real-world examples of reliability to further illustrate his point. Texas has a huge wind-generation system with a capacity of 17,000 megawatts. On August 31, 2016, at noon, output from the wind system fell below 1 percent of capacity. The Texas wind system frequently has swings of thousands of megawatts within a few hours and often produces at less than 5 percent of capacity.

On September 1, 2017, between four and six p.m., California experienced a record demand of a bit more than 50,000 megawatts. By six p.m., two-thirds of solar generation was lost. By eight p.m., all solar generation was lost but demand was still at 46,000 megawatts.

On August 2, 2017, the normally reliable California summer sunshine was interrupted by tropical monsoon weather in Southern California. Solar generation declined by half, and fossil fuel generation had to be mobilized to replace the lost solar.

Rogers emphasizes that wind or solar is an appendage to the electrical grid rather than an essential part of the system. If all the wind and solar vanished, the grid would continue operation without the slightest problem because the grid has to be able to handle the load without wind or solar. Thus, wind or solar does not reduce capital investment for traditional generating plants. Despite reading in the press that coal plants have been replaced by wind or solar, it is never true.

According to Rogers, wind and solar are expensive methods of reducing carbon dioxide emissions. He argues that the only practical method of greatly reducing carbon dioxide emissions from electricity production is replacing fossil fuel with nuclear; wind and solar are a useless waste of money.

Despite these economic realities, Norman Rogers argues ("Green Energy is the Perfect Scam," Rogers 2018e) that the promoters of green energy are not really interested in saving the planet from global warming but in making billions of dollars promoting dumb energy schemes that are nearly useless. What makes their scam extremely clever is that the scammers have convinced the public that the purpose of their scam is to improve the environment and are pretending to be earnest environmental advocates.

For Rogers, any really good scam needs endorsements from authoritative-sounding sources beyond the wind and solar industries that are in on the scam. These sources include nonprofit environmental groups such as the Sierra Club and Greenpeace that need to be seen fighting against an urgent looming catastrophes ("Parade of Impending Catastrophes," Rogers 2018a). Without their scare tactic and catastrophes, no one is going to join their organizations or more important give them money. So the scammers claim that global warming caused by carbon dioxide is the looming catastrophe and that green energy is the solution as it eliminates the burning of fossil fuels, the primary energy source of industrial economies.

Rogers argues that global warming is a magnificent gift to the alarmist science industry as well because it has allowed the climate profession to be corrupted ("The Corruption of Science," Rogers 2014) by pathological science that is focused primarily on increasing the flow of money from governments. Science directed toward discovering truth is out of fashion, and the many scientists ("More Than 1000 International Scientists Dissent Over Man-Made Global Warming Claims," CDepot 2010) who are global warming skeptics don't exist as far as the climate science alarmist industry is concerned.

As well, Rogers includes with the collaborative sources government agencies and politicians who have embraced the threat of climate change by spending billions on subsidizing wind and solar energy. Rogers speculates that the cause gives them something to do that is more noble and romantic than building highways and making trains run on time.

For Rogers, electric utility companies are promoting wind and solar energy ("Dumb Energy Advances in Colorado," Rogers 2018d). Despite these utilities knowing that wind and solar are useless, they are equally aware that because wind- and solar-generated electricity is erratic, it needs to be backed up by reliable conventional electric-generating plants. The only economic benefit is the fuel saved in the backup plants when wind or solar is actually generating electricity. But the cost of the wind or solar electricity is much higher than the benefit of fuel saved. Thus, the utility companies develop business plans that promote the idea that the more wind or solar present in their market, the more money they appear to lose. According to Rogers, since electric utilities are regulated by public utilities commissions, the amount of profit they are allowed is calculated as a fraction of the utilities' capital investment. So the utilities want to make capital investments even if those investments are wind and solar plants that waste money on a grand scale and that results in the electricity consumers' having to bear the cost and the utilities are allowed greater profits.

Rogers then argues that the biggest victim of the green energy scam is the public. Everybody pays more taxes and pays more for energy as a consequence of the scam. However, it is Rogers's position that anyone who criticizes the green energy scam can expect to be ruthlessly attacked. Critics are often depicted as being mental cases as when Gore said that critics of his global warming promotions were like people who thought the moon landing was filmed in a Hollywood studio or thought that Earth is flat. James Hansen, often considered the father of the global warming movement, suggested that executives of fossil fuel companies should be sent to jail for crimes against humanity (Nastu 2008).

Green energy is the perfect scam because it is disguised as a do-good movement and the victims are dispersed, unorganized, and disarmed by liberal left propaganda.

10

Climate Change and the Liberal Big Media

If you tell the truth, you don't have to remember anything. (Mark Train, *Adventures of Huckleberry Finn*)

HAD THE liberal big media acted responsibly, every word spoken at the first major post-Climategate climate colloquium would have indeed built public awareness of the implausibility of man-made global warming and consequently rejected any job-killing legislation, treaties, or regulations designed to control it. But ours is an agenda-driven liberal big media brazenly toting water for the former president and Hill Democrats shamelessly rolling out the BP gulf disaster crash cart to reanimate their flat-lined climate bill.

Although not available online, the *San Diego Union-Tribune* published a column on February 19, 2007, with the title "Fair Reporting on Global Warming" by its media columnist Carol Goodhue, who argued that it would no longer cover both sides of the global warming story and instead limit itself to acting as a conduit for green propaganda.

In "It's official: San Diego paper a propaganda organ" (Fetzer et al. 2007), James H. Fetzer and J. R. Dunn argue there is no drastic change in actual media standards here. The old media has for years acted as a front for various interests ranging from local power centers up to global ideological movements. But it always did so under the pretense of objectivity even when publishing direct handouts from various interests. Media outlets usually took pains to present themselves as objective though the contortions involved often approached the level of caricature.

For Fetzer and Dunn, the aim is obvious. If you can claim that the debate is over for something as shaky as global warming, you can do the same with anything. Abortion? It's over. The existence of Israel? We don't talk about that anymore. The 2017/2018 Trump Russian Collusion? Trump is guilty; no need to wait for Muller's final report. Nothing to it. Once the one-sided news precedent is set, reporters can say any damn thing they please. The word for this is *hubris*.

Fast forward to 2019 and Fetzer's and Dunn's concern about one-sided news reporting and referencing a January 25, 2019, Fox News posting by Joseph A. Wulfsohn, "NBC's Chuck Todd defends not having climate change skeptics on 'Meet the Press' special" (Wulfsohn 2019), in which Wulfsohn provides a recap of NBC News's Chuck Todd defending his decision to not invite any climate change skeptics onto his show during a full-hour special on this subject he was hosting.

In December 2018, Todd dedicated an entire episode of NBC's "Meet the Press" to tackle what he described as the climate crisis and made it clear to his viewers that anyone who questioned the existence of climate change was banned. Todd said, "We're not going to debate climate change, the existence of it. The Earth is getting hotter. And human activity is a major cause, period. We're not going to give time to climate deniers. The science is settled, even if political opinion is not."

Todd appeared January 24, 2019 on *The Daily Show* (Comedy Central/Viacom Media Networks) hosted by Trevor Noah and doubled down on his stance. In the interview, Todd told Noah, "We had a robust debate about taxes. That's the debate. The debate is what do you do [about] this."

Noah asked, "But not for climate change, why not?"

Todd shot back, "Well, I also didn't invite anybody who didn't believe in the moon landing, and I also didn't invite anybody who is flat-earther. Is that alright?"

According to Wulfsohn, Noah continued to press the NBC News political director echoing critics who said, "The press should be giving everyone an equal voice."

Todd responded, "That is—no. Our job is to be fair. Our job is to be fair to the facts that are there. How do we mitigate climate change? How do we—if we're going to build sea walls, where do we build them, and who pays for that?"

In "When climate change warriors can't keep their stories straight" (Joondeph 2017b), Brian C. Joondeph considers the good advice from Mark Twain cited above especially for those who play fast and loose with facts and truth. This is particularly relevant in the internet age when articles, headlines, words, and photos are preserved in perpetuity.

Joondeph warns that lies built upon lies eventually become so tangled that the truth may be forever lost down the rabbit hole. By starting with the truth, one can avoid having to remember the labyrinthine path taken by each additional falsehood.

Joondeph comments that CNN, the network famously referred to by Trump as "fake news," should heed the advice of Mark Twain. Otherwise, they are likely to be tripped up over their own contradictory stories in this case only a few years apart.

Joondeph cites a story ran by CNN in 2015 with the headline "Did climate change cause California drought?" (Sobel 2015). Less than two years later, CNN ran this headline: "California's drought is almost over" (Miller 2017). Joondeph notes the irony of these two CNN articles asking how the drought could be almost over in less than two years. If humans did something to cause it, did they stop doing something that was causing it?

Despite the accusatory Sobel headline tying the drought to climate change, buried in the article is a report from NOAA arguing that the California drought was not due to climate change. In fact, this region suffered megadroughts eons ago, long before humans were driving

SUVs and burning coal for electricity. In other words, the California drought is one of many in this arid region. The reporter makes no effort to research the real causes of the severity of the current drought such as water supply and demand from a growing population in Southern California that is consuming ever-increasing amounts of water and the cyclical nature of droughts.

For Joondeph, those who only scan headlines without reading the entire article see only "California drought" and "climate change" linked. The few who read the article completely recognize the fake news headline for what it is.

In another recent story in CNN with a similarly misleading headline, "Is there a link between climate change and diabetes?" (Howard 2017), Howard has buried in the article the truth that such a link is speculative, an association rather than causation, which contradicts their headline proclamation.

Joondeph argues that speculating via erroneous headlines about spurious associations between unrelated events is dishonest and not scientific. Instead, it's more fake news pervasive in modern journalism. Science deals with causes and effects. While associations are interesting, let science take the next step to determine causation versus association.

Alarmism's constant spewing of fake news diminishes what little credibility remains in the liberal big media. When they do have occasion to report real news, it will be read with the same skepticism as was the boy who cried wolf one too many times.

After Trump withdrew the United States from the Paris Agreement, climate change warriors escalated their "The world is coming to an end" fake-news rhetoric.

In "More Fake News from the Climate Change Warriors" (Joondeph 2017c), Joondeph cites a column in the UK *Guardian*, "World has three years left to stop dangerous climate change, warn experts" (Harvey 2017a) by Fiona Harvey, which was at the time scheduled to be published in the science journal *Nature*. In it, Harvey argues, "The next three years will be crucial." If the Paris Agreement measures are not enacted, even without the US, the world "may be fatally wounded by negligence before 2020." Joondeph notes that Harvey is a former UN climate chief and a member of the IPCC.

Joondeph provides many other examples of similarly scary but failed predictions going back to the high and mighty climate warrior Gore predicting in 2006 (Watts 2006) that "unless drastic measures to reduce greenhouse gases are taken within the next 10 years, the world will reach a point of no return." As of 2019, no such event has been reported, thus proving that Gore's prophesy of "a true planetary emergency" turned out to be a "nothing burger" as Van Jones described (Concha 2017) the incessant blather about Trump and Russia.

Joondeph cites another bastion of fake news, the *New York Times*, which breathlessly predicted in 2014 "The end of snow?" (Fox 2014). Quite certain of the browning of normally snow-covered mountains in winter, they asserted, "This is no longer a scientific debate. It is scientific fact." How did that prediction turn out?

As reported in the *Aspen Times* on May 22, 2017, "Aspen Mountain will open for skiing Memorial Day weekend" (Aspen 2017), Aspen Mountain reopened Memorial Day weekend and commented, "Mother Nature cooperated to make skiing an option" as the reason for reopening. Eighteen inches of new snow the previous week, closer in the year to summer than to winter not only in Colorado but also in California. How about the *Los Angeles Times* reporting on June 20 2017, "Skiers hit the slopes in bikini tops as California's endless winter endures a heat wave" (St. John 2017) that there would be bikini skiing in June this year on "mountains still piled high with snow." So much for the end of snow.

Joondeph then reminds his readers of climate guru and heir to the British throne Prince Charles in early July 2009 being reported in the *Independent* article "Just 96 months to save world, says Prince Charles" (Verkaik 2009). He said to "an audience of industrialists and environmentalists" that "he had calculated that we have just 96 months left to save the world." It's not clear how he made his calculation, but as of

March 2019, we can reasonable assume there are no signs the planet is about to burn up. As a result, it is fair to conclude that his prognostication has turned out to be just fake news.

These are but a very small number of the failed climate predictions promoted by the liberal big media. If the reader is looking for more examples, "The big list of failed climate predictions" (Watts 2014) lists many more failed climate predictions, too many to mention here.

In another twist of liberal left media bias, in "Medical Journals and the Global Warming Noble Lie" (Dunn 2017), Dale Dunn, MD, JD, explains the origin of the term *the noble lie.*

> The Noble Lie is a concept discussed by Plato in the dialogues, lies told by oligarchs to get the populace in the right frame of mind, deceptions intended to influence the mindset and behavior of the populace. The Noble Lie is not often noble; it is the tool of the totalitarian. Totalitarianism is built on the Noble Lie and the best evidence of it in modern society is political correctness and its accompanying censorship and intimidation of any speech or conduct that contradicts the Orwellian "good think" of the Noble Lie.

Dunn explains that he has for many years been collecting research on the effects of warming on human health and counted on the help of Dr. Craig Idso (MS agronomy, PhD geography), who is constantly scanning the scientific literature for research on climate and putting it up on his website CO_2Science.Org in archives of articles. One subject index includes human health. Idso, Dunn, and others have written extensive discussions on warming and human health including chapter 9 in Climate Change Reconsidered (2009; (NIPCC 2009) and chapter 7 in Climate Change Reconsidered II (2013; NIPCC 2014), both published by Heartland Institute of Chicago. Dunn's and Idso's conclusions supported by the medical research around the world on rates of disease and death are that "warming will benefit human health and welfare, for obvious reasons—warm is easier on the plants and animals, so also humans."

Dunn then cites *Lancet*, a multifaceted medical journal entity iconic in medical history that was founded in England in 1823 and now with offices in London, New York, and Beijing; it publishes multiple specialty and general medical journals online and in print. *Lancet* published in 2015 a long-term and planetwide study of death impacts of hot and cold extreme or moderately extreme ambient temperatures ("Mortality risk attributable to high and low ambient temperature: a multicountry observational study," (Gasparrini et al. 2015) by Gasparinni and twenty-two other authors concerning 384 locations around the globe and twenty-seven years of studying 74 million deaths. Their results showed that cold and cooler ambient temperatures killed seventeen times more people than did warmer and hot temperatures.

By contract, Dunn cites an article published on November 1, 2017, by a group created by *Lancet*, "The Lancet Countdown on health and climate change: from 25 years of inaction to a global transformation for public health" (Watts et al. 2017), in which the sixty-four authors produced a fifty-page paper with 195 references declaring a global health crisis due to warming. Dunn asks his readers to consider the contradiction in these two publications by the same medical journal: warm is good, warm is deadly. A cynical Dunn asks,

> Could the *Lancet* editors and the Countdown group they put together be in the bag for the warmer/climate change movement? ... Could this be pushback on their enemy, the warming-skeptical Trump Administration, for leaving the Paris Climate Treaty?

Dunn then challenge most subscribers of prestigious medical journals including the *Journal of the American Medical Association* and the *New England Journal of Medicine* in the US and *Lancet* and *British Medical Journal* in Europe to question how reliable and scientifically trustworthy these publications are and whether they are involved in perpetrating noble lies. Dunn argues that this is simply not so. As a part of academic life and the social structure, medical journals can be counted on to publish junk science that supports Orwellian good think.

Dunn supports his position that the medical academy and its journals are recruited into undertaking junk science to promote the noble lie. To gain this support, Dunn explains that the administrative state needs and thus creates armies of experts to push their agendas with money and rewards of power and position. Funding and research awards and the resulting academic advancements create dependents in academia. Name a liberal leftist cause and without fail, academic medical journals will enthusiastically and cooperatively publish those well-funded research reports and articles in support of the liberal leftist position. Journals are the voice of the academy, and the academy is the mouthpiece of the oligarchic government science advocacy intended to justify government actions.

The noble lie is not restricted to medical journals as described by Dunn. In "The Climate-Industrial Complex" (Rogers 2013d), Norman Rogers cites the endorsements by various scientific organizations that provide much of the muscle promoters of global warming exert. The examples cited by Rogers include the American Geophysical Union, The American Association for the Advancement of Science, The American Meteorological Society, (The Thinking Person's Guide to Climate Change AMS 2019) and National Academy of Science "About the Climate Communications Initiative" (National Academy 2019), which have all officially endorsed the global warming theory. Even groups with no expertise in climate science have piled on with endorsements; these include the American Medical Association ("Confronting health issues of climate change," amednews 2011).

Rogers argues that scientific societies universally claim to be high-minded promoters of science. For example, the American Geophysical Union lists as core values (https//about.agu.org), "the generation and dissemination of scientific knowledge" and "excellence and integrity in everything we do." The reality is very different. Though scientific organizations wear a mask of high-mindedness, they act like lobbying organizations or labor unions. They often form committees to investigate scientific issues. These committees usually come to the same conclusions: We need more money. The government should give us more money. Scientific organizations rarely criticize scientists, scientific projects, or other scientific organizations. Rogers believes that science, also known as big science, is government-supported science. The temptation to influence policy and appropriations by manipulating scientific conclusions is irresistible.

The National Academy of Science released in 2012 a three-hundred-page report, "A National Strategy for Advancing Climate Modeling" (NAS 2012), based on which Roger notes the committee authoring the report was composed entirely of establishment climate scientists mostly engaged in computer modeling of climate. This report paints a glowing picture of climate models and suggests that near-term climate forecasts are possible or will be possible in the future. The reader only needs to check the accuracy of their local televised daily weather forecasts to conclude how that scientific prediction is working out.

Rogers cites the National Academy of Sciences releasing a report in 2012, "Research Universities and the Future of America" (NRC 2012), in which the people responsible for the report were the foxes guarding the hen house, that is, university people. The report modestly suggests doubling the flow of research money from the federal government. Its message polished with an altruistic patina is *Give us more money!*

The American Geophysical Union has a well-organized lobbying effort though it depicts itself as providing information to lawmakers. They run a science policy conference (Rogers 2013b) in Washington, DC, each year. In 2013, the conference featured a workshop on how to lobby members of Congress as well as visits to Capitol Hill to speak with lawmakers.

Former president Eisenhower gave a much quoted farewell address (Eisenhower 1961) in which he warned against the "acquisition of unwarranted influence" by the "military-industrial complex." He had something else to say that is less quoted: "The prospect of domination of the nation's scholars by federal employment, project allocations, and the power of money is ever present—and is gravely to be regarded."

For Rogers, "in holding scientific research and discovery in respect, as we should, we must also be alert to the equal and opposite danger that public policy could itself become the captive of a scientific-technological elite."

Rogers takes the position that Eisenhower's prophecy has come true. Many scholarly fields have greatly benefitted by claiming we are threatened by disastrous global warming, and whole industries have sprung up dependent on government funding and regulation. There is plenty of pushback from individual scientists ("More Than 1000 International Scientists Dissent Over Man-Made Global Warming Claims," Morano 2010), but this climate industrial complex sticks together and does its best to marginalize the critics.

Richard Lindzen, a pioneer in climate science and a professor at MIT, wrote about the politics of climate science in "Climate Science: Is it currently designed to answer questions?" (Lindzen 2009). In it, he traces the history of government support of scientific research and describes how scientific seekers of government money adopted the promotion of fear as an incentive to open the floodgates of government money. Some of the fears he catalogues are the fear of cancer, fear of the Soviet Union, and fear of national decline.

The fear of catastrophic climate change is working well for many scientists and not just climate scientists. Lindzen describes how activists gain influence and power in scientific organizations. He suggests that when there is a conflict between models (theory) and data, there is a great tendency in the climate research community to change the data rather than revise the theory. This can often be done by reexamining the data and adjusting the analysis methods.

Rogers argues the enforcers of the climate warming dogma will attack those who gain traction in rebutting global warming dogma. He points to their strategy of enforcing political correctness, that is, the belief that certain ideas and thoughts are socially unacceptable and that those who utter these ideas or thoughts risk becoming socially unacceptable. Unless they are exceptionally courageous, scientists keep quiet because they don't want to be run over by the enforcers of climate correctness.

Rogers cites a famous attack on a dissenter, Dr. Bjorn Lomborg, in January 2002 by *Scientific American*. Lomborg wrote *The Skeptical Environmentalist: Measuring the Real State of the World* (Lomborg 2001) that questions much of the environmental dogma including the global warming dogma. In a mystifying overreaction, *Scientific American* devoted eleven pages to trashing Lomborg and his book. Among the critics contributing to the attack was John Holdren, Obama's former science advisor. When Lomborg tried to answer the attack on his website ("Bjørn Lomborg's comments to the 11-page critique in January 2002 Scientific American (SA)" (Lomborg 2002), *Scientific American* at first threatened to sue him for copyright violation because he quoted text from the attacks. In the end, the campaign to destroy Lomborg failed and instead made Lomborg famous.

Rogers emphasizes that scientific organizations that endorse global warming catastrophe theory are really just fancied-up labor unions and their reports and statements are generally self-serving declarations disguised as objective analysis. It is obviously foolish to ask scientific organizations to give objective advice concerning programs in which they have deep self-interest. The National Academy of Sciences says in its mission statement (NAS 2019) that it is to give "independent, objective advice to the nation on matters related to science and technology." The problem is obvious. The government should seek out persons and organizations without a self-interest stake when asking for advice concerning science policy and science spending.

In providing an explanation for why the scientific community can justify corrupting its scientific integrity, James Lewis in "$cience Mag Jumps on Global Moneywagon" (Lewis 2007b) states, "Scientists like money." Big science is a big business supporting nearly half the budgets of major universities. Science professors are hired only if they can swing enough federal grant money to pay for their labs, hire a gaggle of graduate assistants, and let the universities skim up to 40 percent for overhead. And besides, it's nice to get fat salaries. So the professional scientists' union, the American Association for the Advancement of Science, has ads headed *"AAA$"*; it isn't shy about it.

Lewis argues the trouble here is that money means politics, and politics means shading the truth. This results in politicized science, which corrupts real science and makes it very hard to publish contrarian science studies that do not conform to the politically correct science of the day.

In a December 14, 2007, article in *Science* magazine, "Year of the Reef" (Kennedy 2007), Donald Kennedy provides an example of how science can be ruined when he describes how coral reefs are dying. It's like the *National Enquirer.* In this article, Kennedy states, "The United States could … mitigate carbon dioxide emissions: The root cause of global warming and the reef problem. *Experience suggests that for this, we might have to await an election."* Though *Science* was not exactly endorsing a Democrat candidate for president at the time, it is like the union boss telling members how to vote in a general election if they want to get more money.

As science fiction guru Arthur C. Clarke loved to point out, famous physicists predicted in the early 1900s that man would never fly. In the 1950s, they confidently said that a moon landing was impossible. "Clarke's Law" states (Wikipedia, Clarke) that whenever a famous scientist tells you that something is impossible, don't believe it; chances are they are just wrong. This same advice is even truer for predictions being reported in the liberal big media.

Humans are the fastest-learning creatures ever known. In the last hundred years, we have gone from choo-choo trains to scramjets. Give us another century and who knows what we will do? Colonize Mars? Solarize energy? Double our life span? Human history gives lots of grounds for hope and much less for despair.

Lewis worries whether politicized science can ever be fixed. If we allow the search for truth to be so easily twisted by political fads, we may be in really deep trouble.

Climate skeptics are frustrated by the liberal big media chasing and promoting bogus stories such as the Trump-Russia collusion when this same media could be asking the climate change alarmists who continue to predict imminent apocalypse to explain their past failed predictions.

The reasons they don't are quite obvious; the liberal big media are committed to promoting their liberal leftist agenda whether it is man-made global warming, Russia hacking the election, or getting Democrats elected to the House, Senate, or presidency of the US. For lazy editors and money-hungry owners of the media outlets, these professional and business goals can be easily achieved simply by accepting at face value a corrupted UN official asserting that the world is doomed unless the US throws trillions of dollars at a mythical problem.

It is becoming obvious to those who care about maintaining their integrity that journalists don't really care about honesty and good reporting and instead pursue ratings and social media followers with their latest outrageous attempts to advance the liberal left agenda that can't be sold to voters.

11

Demonizing Skeptics of Climate Change

In "Is There Anything Climate Change Can't Do?" (Joondeph 2017a), Brian C. Joondeph mocks the notion that daily news stories focus on all the things Trump can do wrong. Trump causes terror attacks. Trump causes nightmares (Hunter 2017), insomnia, and eating binges. Trump caused Barbara Streisand to gain weight. Trump is omnipotent. Yet he has competition. Not Obama. Not the deep state. Instead, global warming, now known as climate change and now being framed as causing all severe weather is the new supervillain on the block.

According to Bernie Sanders, climate change causes terrorism. Just like Trump, climate change also leads to severe weather—heat, cold, rain, snow, floods, and droughts—you name it. Just when you thought the climate change superhero could do it all, it outdid itself by causing Brexit and diabetes as declared by global warming oracle Gore speaking at an event where he announced the long-anticipated sequel to his 2006 blockbuster hit *An Inconvenient Truth*. The same movie predicted an environmental apocalypse (French 2016) within ten years, meaning 2016, which as of 2019 has yet to occur.

In "Climate change helped cause Brexit, says Al Gore" (Johnson 2017), Ian Johnson quotes the former vice president telling his audience, "Brexit was caused in part by climate change" and notes that extreme weather is creating political instability "the world will find extremely difficult to deal with." Joondeph takes this prediction of political instability as a new phenomenon, not a recurrent theme in human civilization since the days when humans began to govern themselves.

Gore further links the Syrian drought to a refugee influx into the UK and the 2016 Brexit vote. Anyone with a shred of common sense would conclude these same droughts would have affected neighboring countries such as Israel, Jordan, and Turkey and caused those countries to send refugees to Europe. Gore's narrative has a number of large holes, but when it comes to climate change, why bother with inconvenient truths?

On this point, Joondeph argues that there is a veritable online cottage industry cataloguing hysterical, failed predictions of environmentalist catastrophes. At the American Enterprise Institute, Mark Perry keeps his list of "18 spectacularly wrong apocalyptic predictions" (Perry 2015) made around the original Earth Day in 1970. Robert Tracinski at *The Federalist* has a list of "Seven big failed environmentalist predictions" (Tracinski 2015). The *Daily Caller's* "25 years of predicting the global warming 'tipping point'" (Bastasch 2015) makes for amusing reading including one declaration that we had mere "hours to act" to "avert a slow-motion tsunami." This hysteria has found its way into the economic and political institutions of free societies of the West and is treating the freedoms its citizens enjoy by maligning those skeptics who question the faith of the climate alarmists.

Noel Sheppard's "Weapons of Global Warming Destruction" (Sheppard 2007) argues that the center of the battle between socialism and capitalism is clearly man's "undeniable" role in global warming and the lengths the liberal left will go to advance their forces on this issue including threats to people's jobs if they don't conform to the views expressed by liberal leftists.

As an example, Sheppard notes that in December 2006, The Weather Channel's foremost climate expert Heidi Cullen called for meteorologists to be stripped of their certifications (Neudorff 2007) by the American Meteorological Society if they disagreed with the tenets of AGW.

In another example from early 2007, Sheppard cites that the Democrat governor of Oregon threatened to remove (Patton 2007) the mostly ceremonial title of State Climatologist from a man who, since 1991, had questioned man's role in climate change.

On February 7, 2007, Gore told a group he was speaking in front of in Madrid, Spain (Gore 2007), that China was correct to blame America for global warming. Maybe more important, Gore stated to the crowd that this Asian country—which happens to possess the fastest growing economy on the planet with the largest population and is the world's largest emitters of air pollution—should not be required to participate in climate solutions until the United States does its share.

According to Sheppard, those familiar with the strategies employed by socialists must recognize the common tactic of blaming the world's problems on America while absolving all other nations of any responsibility.

In reality, by exaggerating the dangers of global warming as well as humanity's real or imagined role, Gore and his ilk are using exactly the same fear tactics they claim the Bush administration was guilty of in the lead-up to the March 2003 invasion of Iraq. Sheppard argues that somehow, the media have deliberately overlooked this extraordinarily delicious hypocrisy as they themselves fall hook, line, and sinker for the conclusions they so desire to hear from those they hold in highest esteem.

Expanding on the common liberal left tactics for eliminating debate on any climate change issue by skeptics is the tactic of casting them as vile vermin as evidenced in a February 9, 2007 *Boston Globe* article by liberal writer Ellen Goodman, who offensively proclaimed in "No change in political climate" (Goodman 2007), "I would like to say we're at a point where global warming is impossible to deny. Let's just say that global warming deniers are now on a par with Holocaust deniers, though one denies the past and the other denies the present and future." In "Global Warmists Exploit Holocaust" (Sheppard 2007), Marc Sheppard argues that when Goodman likened climate skeptics to holocaust deniers in that article, she raised more than a few eyebrows. Yet hers was not the first reprehensible use of that fetid analogy or unfortunately the last. In truth, environmentalists' deplorable trivialization of Hitler's genocide can be traced as far back as 1989, when Gore wrote a scare piece for the *New York Times* under the improbable title "An Ecological Kristallnacht. Listen" (Gore 1989). Predicting a 5 degree C rise in global temperatures "in our lifetimes," he warned of nightmares fifty years past.

> In 1939, as clouds of war gathered over Europe, many refused to recognize what was about to happen. No one could imagine a Holocaust, even after shattered glass had filled the streets on Kristallnacht. World leaders waffled and waited, hoping that Hitler was not what he seemed, that world war could be avoided. Later, when aerial photographs revealed death camps, many pretended not to see. Even now, many fail to acknowledge that our victory was not only over Nazism but also over dark forces deep within us.

Sheppard schools his readers with the details of the historical event referred to as *Kristallnacht*, German for "Crystal Night." The very name elicits lurid images of that dark night in November 1938 when Germans throughout the land were awakened to the sights and smells of burning synagogues, the noise of shattering window glass, and the screams of innocent Jews being savagely beaten. It was a night when thousands of Germans who read the signal that Jews were *Vogelfrei*, "fair game," joined Hitler's *Sturmabteilung* (brown shirts) in killing at least a hundred and dragging thirty thousand more men, women, and children one step closer to the death and agony awaiting them in Nazi concentration camps. Sheppard acknowledges that Gore's repulsion mission had been accomplished.

Gore then dared compare the world's failure to respond then to contemporary environmental complacency and "dark" self-interest:

> In 1989, clouds of a different sort signal an environmental holocaust without precedent. Once again, world leaders waffle, hoping the danger will dissipate. Yet today the evidence is as clear as the sounds of glass shattering in Berlin.

Sheppard was appalled by the audacity of recalling the very sounds that evoked the tag *Kristallnacht* and suggests that those disregarding Gore's personal delusions of skeptics' "self-destructive behavior and environmental vandalism" were somehow synonymous with a "world [that] closed its eyes as Hitler marched" and betrays a mind at once deluded and devious.

When the same words appeared in Gore's 1992 *Earth in the Balance: Ecology and the Human Spirit*, Matthew Brooks, executive director of the Republican Jewish Coalition, noted his outrage:

> For the vice president to equate the utter horror and the tremendous tragedy of Kristallnacht to try and invoke the passion of people about the environment is an insult to all the people who were victims of the Holocaust.

Sheppard then makes reference to Deborah Tannen, who on March 27, 1998, appeared on Jim Lehrer's NewsHour (Gergen 1998) to discuss her *The Argument Culture: Moving from Dialogue to Debate*. After declaring global warming settled fact and skeptics as oil-company shills, the author took Gore's analogy to its next illogical step by listing global warming skeptics right alongside Holocaust deniers.

Sheppard further argues that the word *holocaust* itself is a weapon being exerted by climate scaremongers in much the same manner that the word *racist* has been wielded by racemongers to shut down debate by questioning the opposition's morality. If you were against the first Holocaust, they attest, you must be equally against the coming climate holocaust; otherwise, you are undoubtedly immoral.

It was a full eight years before the liberal big media would advocate Gore's "debate is over" mantra in response to the junk-science he spewed in *An Inconvenient Truth* that the reviled global- warming denier was born.

Sheppard argues that with the Holocaust-level guilt established in the psyches of the indoctrinated, the next step was to further marginalize heretics by projecting imaginary Holocaust-worthy punishment. In doing so, in true authoritative fashion, freedom of speech and any open, rational debate on the subject of climate change was effectively shut down.

In September 2006, *Grist* magazine's equally hysterical David Roberts suggested (Roberts 2006) that the method by which deniers should be held accountable for the consequences of their inactions should mirror that by which the most heinous of Nazi war criminals had been for their actions:

> When we've finally gotten serious about global warming, when the impacts are really hitting us and we're in a full worldwide scramble to minimize the damage, we should have war crimes trials for these bastards—some sort of climate Nuremberg.

Sheppard notes Roberts's outrageously moronic suggestion that those offering dissenting scientific opinion on the causes, scope, dangers, and anthropogenic influence of global warming are guilty of similar crimes against humanity is nothing short of madness.

Sheppard reminds that on October 22, 2006, James Hansen, former director of NASA/ GISS gave a testimonial before the Iowa Utilities Board (Hansen 2006). His appearance represented but one battle in his ongoing war against commissioning any new coal-burning power plants prior to the development and deployment of successful carbon dioxide sequestration technologies. Amid the requisite dooms-day floods and extinctions, Hansen avowed,

> Coal will determine whether we continue to increase climate change or slow the human impact. Increased fossil fuel CO_2 in the air today, compared to the pre-industrial atmosphere, is due 50 percent to coal, 35 percent to oil and 15 percent to gas. As oil resources peak, coal will determine future CO_2 levels. Recently, after giving a high school commencement talk in my hometown, Denison, Iowa, I drove from Denison to Dunlap, where my parents are buried. For most of 20 miles there were trains parked, engine to caboose, half of the cars being filled with coal. If we cannot stop the building of more coal-fired power plants, those coal trains will be death trains - no less gruesome than if they were boxcars headed to crematoria, loaded with uncountable irreplaceable species. Sheppard interprets Hansen's words to be clearly crafted to evoke vivid imagery of the cattle cars used by the Nazis to transport European Jews to camps in Poland built for the purpose of their extermination and no other. Such cars were the rolling counterpart of death marches in which entire families were forcefully packed standing room only with little light or ventilation and no latrines, food, water, or hope; they had to endure torturous journeys often lasting several days and rarely ending in anything short of their cremation.

To call out the Hansens and Gores of the world and their despicable conflation of climate change skeptics to Holocaust deniers, it is necessary to provide further context as to how the climate change doomsday movement has been sustained for decades by citing specific incidences that help expose the corrupt campaigns used to smear climate skeptics.

In "Gore's RICO-style prosecution of Global Warming Skeptics" (Cook 2016), Russell Cook cites a March 29, 2016 press conference (Schneiderman 2016) led by New York State Attorney General Eric Schneiderman, who was accompanied by his fellow attorney generals and Al Gore. At the press conference, Schneiderman announced the latest effort to use racketeering laws to prosecute "climate change deniers.- Minutes after first thanking Gore for attending and noting how his 2006 movie *An Inconvenient Truth* galvanized the world's attention on the urgency to act on climate change, Schneiderman spoke of how the group was working to find creative ways to "enforce laws being flouted by the fossil fuel industry and their allies in their short-sighted efforts to put profits above the interests of the American people."

"We know what's happening to the planet," Schneiderman continued. "There is no dispute, but there is confusion, sowed by those with an interest in profiting from the confusion and creating misperceptions in the eyes of the American public."

Gore was allowed to offer a mini variation of his movie talk halfway through the press conference; he then injected himself into the Q&A session at the end regarding a report from the *New York Times* (Gillis et al. 2015) that indicated that Exxon, unlike tobacco companies, had "published extensive research over decades that largely lined up with mainstream climatology" thus contradicting the alarmists' analogy about deceptions from Big Oil and Big Tobacco.

Gore's answered this question by stating,

> I do think the analogy may well hold up rather precisely ... Indeed, the evidence indicates that these journalists collected, including the distinguished historian of science at Harvard Naomi Oreskes, who wrote the book *The Merchants of Doubt*, that they hired several of the very same public relations agents that had perfected this fraudulent and deceitful craft working for the tobacco companies.

Cook argues that both the nearly one-hour press conference and Gore's ninety-six-minute-long movie boil down to just two claims: that the science is settled and that naysayers are corrupted by fossil fuel funding; there is thus supposedly no need for the public or journalists to listen to any skeptics because of claims one and two.

Cook identifies the source for this advice; it ultimately stems from author Ross Gelbspan, whose 1997 *The Heat is On* (Gelbspan 1997) is routinely cited regarding the accusation that skeptical climate scientists are tainted by fossil fuel industry funding. Gelbspan says they operate under a sinister coal industry leaked memo directive to "reposition global warming as theory rather than fact," a purported replay of old tobacco industry maneuvers. Gore made that exact comparison in his 2006 *An Inconvenient Truth* movie in which he spelled out that same memo phrase to insinuate a quid pro quo association where skeptic climate scientists were paid to follow the orders of "big coal and oil." The problem with this accusation for Cook is that neither Gore nor Gelbspan nor anyone else who repeated the accusation ever bothered to provide their audience with full-context document scans, under-cover video/audio transcripts, leaked emails, money-transfer receipts, or other actual physical evidence to prove that such a blatantly corrupt association exists.

Cook did find and read the full context of the coal industry leaked "reposition global warming" memos Gore and Gelbspan claim are smoking-gun evidence of skeptic climate scientists' guilt. Cook notes the memos were part of a tiny, short-lived pilot project PR campaign that practically none of the public saw. Worst of all for Gore, his companion book for his movie said (Gelbspanfiles 1997) that the Pulitzer-winning reporter Gelbspan discovered the leaked memos. Yet Gore's 1992 *Earth in the Balance* book quoted (Gore 1992) from those same memos years before Gelbspan first mentioned them (Curwood 1995). And Gelbspan never won (Gelbspan 2013) a Pulitzer.

For Cook, the irony of the crimes of sowing public confusion into the climate change scam by Schneiderman and Gore in their press conference was completely reversed on who sows the confusion and who should be facing legal action. The single highly questionable phrase in Gore's movie cited above should have sunk this racketeering investigation effort and Gore's entire legacy.

In a second article, "Smearing Climate Skeptics" (Cook 2014), Cook cites a February 28, 2014, *Wall Street Journal* article "Jenkins: Personal Score-Settling Is the New Climate Agenda" (Jenkins 2014) in which Holman W. Jenkins Jr. started out with a great recitation of efforts to silence and marginalize critics of man-made global warming skeptics describing their "indications of a political movement turned to defending its self-image as its cause goes down the drain."

For Cook, the climate change crisis was arguably dead on arrival from the early 1990s when skeptical climate scientists offered science-based criticisms that apparently fatally undermined those of the UN IPCC, which Gore unquestioningly endorsed. It should have been no surprise to anyone that Gore and his enviroactivist friends felt a need to marginalize climate scientist skeptics in the eyes of the public to the point that nobody would ever give them the time of day. Gore and friends ultimately boiled their message down for the liberal big media and the vulnerable public to an easily memorized three-point mantra around 1996: "The science is settled ... Skeptic scientists are industry-corrupted ... Reporters may ignore skeptics for the prior two reasons."

For readers asking, "What skeptical scientists?" the brilliance and effectiveness of that last talking point shines through. Despite widespread cries from enviroactivists and reporters that skeptics are given unwarranted attention, Cook challenges his reader to ask themselves when the last time was that they saw a global warming news report in which skeptical climate assessments were thoroughly spelled out and how many times they had seen that done in the decades-old history of this issue.

In answering those two questions, Cook undertook an ongoing detailed count (Cook 2013) of America's most respected TV news outlet, PBS's NewsHour's online broadcast transcripts from 1996 to 2014, which revealed that it had around four hundred times discussed or at least mentioned in a significant way the global warming issue on the program and that viewers were given skeptic science points only five times, four of which were fleeting enough to be easily missed. Not much of an objective approach given this critical issue.

For Cook and others, this situation no longer looks like any kind of credible claim that skeptical climate scientists were manufacturing doubt about the science at the behest of the fossil fuel industry; it instead takes on the appearance of a concerted effort to manufacture doubt about any critics of man-made global warming.

Cook concludes that as the so-called global warming crisis looks ever more as if it's headed toward certain collapse, tough scrutiny of all the efforts to erase skeptic climate scientists from the public eye might end up revealing the most inconvenient truth of all: that the global warming issue has remained alive all this time through constant questionable infusions of information about its science claims and through unsupportable character assassination of its critics.

If the idea of man-made global warming is exposed to have been faulty from the start, it could end up as the biggest ideology collapse of all time.

12

The Hypocrisy of Climate Alarmists

VACLAV KLAUS, who served as president of the Czech Republic until
2013 and lived under Soviet tyranny, knows firsthand the characteristics of totalitarians. In "Climate alarmists pose real threat to freedom"
(Klaus 2008), Klaus warns climate alarmists of their hypocrisy when
they call for humankind to sacrifice economic activity for the sake
of reducing carbon dioxide emissions while indulging themselves in
large carbon footprints generated by their mansions, SUVs, and private jets. In "Climate totalitarians" (Schmitt 2008b), Jerome J. Schmitt
is reminded of George Orwell's *Animal Farm: A Fairy Story* (Orwell
1945), an allegory for the Soviet revolution wherein the pig commissars
abridged the farm's commandments to state, "All animals are equal but
some animals are more equal than others."

Ethel C. Fenig supports this observation in "Two houses symbolize two Americas" (Fenig 2007). She concludes that there are apparently two Americas with a reference to Jeff Dobbs's article "The two
Americas of John Edwards" (Dobbs 2007). One is the lavish lifestyle
of the Democrats who are willing to tax others so they can live the
Democratic way while the other is low keyed and unassuming earlier
known as cloth-coat Republicans.

As befits the rewards of a former senator, a former vice president
of the US, a former presidential candidate, and an Academy Award
winner, Gore, who claims to be a deeply concerned environmentalist

and climate change fighter as revealed in the web posting "A Tale of Two Houses" (Mikkelson 2008), lives in a twenty-room mansion (not counting eight bathrooms) heated by natural gas. Add to that a pool (and pool house) and a separate guest house all heated by gas. In one month, this residence consumes more energy than the average American household does in a year. The average bill for electricity and natural gas runs over $2,400 per month. This property consumes more than twenty times more natural gas than does the average American home. Gore's Nashville home is "approximately four times the size of the average new American home" and consumes electricity at a rate of about 12 times the average for a typical house in that Southern city with a relatively mild climate.

According to Fenig, Gore defends his excessive carbon footprint by his purchases of carbon offsets from someone or something with an apparently tiny paw print to justify his hypocrisy.

Perhaps Gore could have learned from another wealthy and successful individual in the South cited in the "Two Houses" article who quietly and without fanfare lives in a home incorporating every green feature current home construction can provide. The house is four thousand square feet (four bedrooms) and is nestled on an arid, high prairie in the American southwest. A central closet in the house holds geothermal heat pumps drawing groundwater through pipes sunk three hundred feet into the ground. The water (usually 67 degrees F) heats the house in the winter and cools it in the summer.

The system uses no fossil fuels such as oil or natural gas, and it consumes a quarter of the electricity required for a conventional heating/cooling system. Rainwater from the roof is collected and funneled into a 25,000-gallon underground cistern. Wastewater from showers, sinks, and toilets goes into underground purifying tanks and then into the cistern. The collected water then irrigates the land surrounding the house. Surrounding flowers and shrubs native to the area enable the property to blend into the surrounding rural landscape.

Located in Crawford, Texas, and dubbed the Texas White House by the press, this is the home of George W. and Laura Bush.

This hypocrisy is particularly evident in the ruling class of a society. In "Global Warming: Making the Ruling Class into the Crackpot Class" (Rogers 2015b), Norman Rogers defines the ruling class from which the likes of Gore emerge as being privileged, wealthy, well educated, and having important jobs. This is the class from which many important leaders are drawn. According to Rogers, the problem with the ruling classes over time as theorized by the Italian sociologist Pareto (Rogers 2015b) is they go soft and lose their vigor and sense of purpose over time and are replaced by tougher upward strivers.

To make his point, Rogers compares two secretaries of state. John Foster Dulles (Dulles 1998) was born in 1888 and was Eisenhower's secretary of state during the 1950s. John Kerry was born in 1943 and was Obama's secretary of state. Both of these men were born into the ruling class.

Dulles had an uncle and a grandfather who were secretaries of state. Dulles entered Princeton at age sixteen and graduated as valedictorian. Dulles distinguished himself in a long career in government service and as an international lawyer. He organized countries of the world to resist Communist expansionism during his term as secretary.

John Kerry's maternal relatives were members of wealthy old-money families (Wikipedia, Kerry), and he is the beneficiary of various trusts from these relatives. He went to elite New England prep schools and Yale, where he was a mediocre student (AP 2005). He has been married twice, both wives coming from wealthy circumstances.

Rogers describes Kerry as tall and impressive looking. He is a talented politician but lacks intellectual depth. According to Kerry, climate change is the most serious problem (IBD 2014) the world faces. He wasn't much worried about international relations, though Rogers notes that the Obama administration ran its foreign policy out of the White House granting little discretion to the secretary of state.

For Kerry, a vociferous belief in the danger of climate change may be a marker for a shallow thinker. It doesn't take a lot of research to realize that his story "We are doomed by climate change" is full of holes.

As another example of the ruling class's shallow view of the world, Rogers cites Adele Simmons, one of the bluest of Chicago old-money bluebloods. Her ancestors were accomplished and wealthy. She was the president of Hampshire College in Massachusetts. She left Hampshire to become president of the John D. and Catherine T. MacArthur Foundation, one of the richest foundations in the country. That foundation gives out genius awards often to liberal left recipients. An acquaintance of Rogers who attended Hampshire and works with foundations said, "She destroyed both Hampshire and MacArthur by taking them to the far, far left ... I think her a cultural quisling of the first degree." Simmons has since left the MacArthur Foundation.

Rogers goes on to characterize Simmons as a cheerleader for the gimmicks beloved by the climate doom crowd—plenty of studies, pronouncements, windmills, solar power, and so on. According to Simmons, when a couple of minor coal-fired plants were closed in Chicago, it was a big victory made more so because one plant was in a Hispanic neighborhood. Even if you were to take seriously the predictions from the computer climate models, closing a coal plant in Chicago is of no importance when a new and bigger plant is being built in China every week. As a member of the climate change promoters, Simmons is engaged in feel-good symbolic activities—nothing more. But these promoters take themselves very seriously.

Rogers cites as another example of a member of the ruling class Tom Steyer, a wealthy climate activist in San Francisco. His father was a partner in the law firm of Sullivan and Cromwell, John Foster Dulles's old firm. His wife, Kathryn (Kat) Taylor, is the granddaughter (Taylor 1986) of the former president of the Crocker National Bank. Steyer went to Yale, and his wife went to Harvard. They have professional degrees from Stanford. Steyer made more than a billion dollars as chairman of the investment firm Farallon Capital. Ironically, his firm invested heavily in coal (Hinderaker 2014). He may not have had the global warming religion back then.

According to Rogers,

> Steyer is like a slightly daffy English nobleman who discovers socialism and pesters his servants, trying to understand their lives. Steyer even has the modern equivalent of a noble estate – the 1,800-acre TomKat Ranch on the California coast. There he raises Leftcoast Grassfed (Leftcoast 2019) beef.

> These members of the ruling class had connections and positions. They were rooted. They had a sense of belonging and a sense of responsibility. However, they lost confidence in their class worthiness. Rather than ignoring crackpots, they became crackpots. John Kerry, Adele Simmons, and Tom Steyer being born into the ruling class and rather than becoming serious members of that class, they took up outré political activism – in their case, global warming.

Rogers argues,

> that crackpots and eccentrics have always been part of the privileged classes. But no one took them seriously, least of all the members of their same class. Now the crackpots are in high positions and are being taken seriously by the liberal big media. In some regions, like California, it's gotten to the point that common sense has been turned upside-down, and unless you believe in fashionable crackpot causes like global warming, you *are* the crackpot.

In another example of environmental hypocrisy in the ruling class, Rick Moran, in "Rank hypocrisy from George Soros" (Moran 2008), cites liberal financier and backer of far-left groups like Moveon.org

George Soros as having been caught in one of those "Do as I say, not as I do" moments. Moran explains that Soros, a huge proponent of the global warming theory, has purchased an $800 million stake in a Brazilian oil company. According to the *Wall Street Journal,*

> But as one of the world's most successful investors and speculators, Mr. Soros seems to have a different set of standards when it comes to his own wallet. His moral indignation takes a back seat to the profit motive.
>
> To wit, the billionaire do-gooder has purchased an $811 million stake in Petrobras, "making the Brazilian state-controlled oil company," according to Bloomberg News on Friday, "his investment fund's largest holding."
>
> It's not only that Petrobras is a fossil fuel company. The more interesting aspect of the Soros investment is that Mr. Soros's Petrobras investment cannot be profitable if the company does not exploit its Tupi oil field, the largest offshore find in the hemisphere. Indeed, Petrobras is rapidly emerging as a world leader in technology to exploit such no-no reserves, while Brazil has thousands of miles of pristine coastline and a large indigenous population. Is Mr. Soros not outraged that he will be funding corporate interests that threaten these? Apparently not as long as there is money to be made that will go into his own pocket.

Soros certainly knows about offshore investing. His hedge fund is registered and domiciled in an offshore tax haven therefore obscuring its investors and investments alike.

Frustrated climate change promoters often demand that those who are skeptical be prosecuted and jailed (Weinstein 2014). That should be enough to convince any sensible person that something is rotten in the climate change establishment. A lot *is* rotten. Predictions of climate doom come from computer models that are absurdly inadequate (Rogers 2014) and heavily manipulated. Junk science rules because junk science has vastly improved the career prospects of the scientist promoters of climate doom. For environmental organizations, climate change is the latest scare story and useful for fundraising.

In "Global Whining" (Kaufman 2007), Ari Kaufman cites a quote by Gore on February 3, 2007, gloating over the forty-six nations, most of them European, that backed a new French-led environmental body to save us from global warming: "We are at a tipping point. We must act, and act swiftly ... Such action requires international cooperation."

Kaufman sarcastically comments that even back in 2007, unless stuck in a cave or flying around in one of Laurie David's private planes, one couldn't help but hear that "warming" is approaching us faster than ever.

And then the hypocrisy—the most fervent critics of industrial policies are typically the folks who are far too important to practice what they preach. Kaufman argues that "enviros" such as Laurie David, Ted Kennedy, and Barbra Streisand will whine and kvetch until they can hate America no more, but ask them to close down the 12,000 square foot air-conditioned barns at their Malibu estates or install electricity-generating wind turbines along their sailing route on Nantucket Sound and one can expect a tongue-lashing. John Edwards, the populist who incessantly rants about "Two Americas," had a 30,000 square foot house built for him. Apparently, this is more of that "Do as I say, not as I do" liberal doublespeak that Michael Moore employs while he flies his jets to his mansions financed with Halliburton stock.

Kaufman reminds his reader that at the end of Gore's award-winning film *An Inconvenient Truth*, Gore jumps into his private jet. Kaufman hopes sarcastically the jet is using "environmentally sensitive fuel."

In "A Modest Proposal to Eco-Celebs" (Feldman 2007), Clarice Feldman tries to figure out why Norman Lear with a twenty-six-car garage insists that everyone else should be cutting his or her driving and why Barbra Streisand from her well-staffed mansion overlooking the Pacific Ocean advises her fans to air dry their laundry outside. Teresa Heinz and John Kerry, who use a private plane to travel to their five mansions and SUVs, warn everyone to cut back on energy use.

Feldman believes he has figured out what this naked hypocrisy is about. It's not just scientific and economic illiteracy on their part; it's a narcissistic desire to widen even further the gulf between themselves and those beneath them on the economic and social ladder while clothing their desires in some moral purpose. This is nothing new of course. Feldman notes that at various times and places throughout the world, what one wore including colors and fabrics, the length of swords, and how much the tips of shoes could curl were set by law to make sure no one mistook the milkmaid and yeoman for the lord and lady.

Feldman argues that because ample energy supplies are critical and that continued economic growth is vital to the health of the nation and the only significant way to help the poor and starving of the world, this movement if successful would add to the tragedies already wrought by the combination of stupid celebrities and scientific ignorance. They spearheaded the destruction of the nation's nuclear power industry and the immiseration and death of people throughout the world by the banning of DDT to fight malaria. Naturally, these ecowarriors create narratives in the corrupt liberal big media to help us forget altogether what listening to their idiocy has already cost us.

13

How Climate Science
Research Should Work

To ADDRESS the current climate warming and climate change hysteria being forced down our intellectual throats, let us step back and understand how scientific breakthroughs actually find their way into mainstream acceptance, commercialization, and scientific and public consensus.

In "Who's Afraid of Global Warming?" (Dunn 2007), J. R. Dunn argues,

> Science works by means of prediction. Once data is collected and evaluated, and a hypothesis formed, scientific method requires that certain predictions be made to act as tests of the overall theory. If the predictions work out, we can regard the hypothesis as proven. If not, scientists vow to do better next time.

Dunn says there exists today a large and growing class of theories on subjects that are too vast, too small, too remote, or too complex to allow adequate testing. These include such abstruse concepts as string theory, brane theory, and dark matter. So critical has the situation become in some fields that there has been talk of the end of physics or even the end of science as a whole. Dunn considers that a premature diagnosis.

For Dunn and many other scientists, one of the familiar cutting-edge ideas not susceptible to testing is global warming. Dunn is of the opinion that the earth's climate is far too large and complex in terms of mathematical and common meanings to be subject to any conceivable form of testing. Yet we are being told by the climate alarmists that the dangers presented by climate change are so great that we cannot wait for actual evidence. The risk is infinite, so we have to act now while there's still time.

To challenge these dire predictions, Dunn concludes that given that the earth's climate system exceeds any testing using incomplete computer modeling, it should be possible to get a better understanding of the earth's current climate reality by looking at previous climate cycles and from this understanding develop theories that project what can be expected in the future.

One such climate event took place in relatively recent historic time, the Little Climatic Optimum (LCO), better known as the Medieval Warm Period. During the LCO, worldwide temperatures rose by 1 to 3 degrees C for roughly three hundred years beginning in the tenth century and ending late in the thirteenth century. Cooling occurred again during the Little Ice Age, which ended around the 1840s, and it has been warming since. Records from these three eras are abundant and easily available.

For Dunn, however, before any predictions can be made on future climate change cycles, there needs to be agreement in the climate research community of scientists as to the data set(s) that best represents the extremes plotted for each of the identified cycles above. Dunn asks whether scientists can determine whether the earth is cooling or warming without a basic understanding and agreement on the past history of the earth's climate cycles. Do humans and fossil fuels have nothing, or something, or everything to do with it?

To establish a sound scientific basis for answering these and other questions related to Earth's climate cycles, Howard Hyde ("Climate Change: Where is the Science?" Hyde 2015) argues,

> The only way to find out is through the most rigorous and critical application of the scientific method, from laboratory practice to public discourse. Anything less than that increases the risk that the "solution" could be more catastrophic to humans than the results of climate change itself.

Hyde then argues that nothing defines science better in the popular mind than the predictive power of scientific theory.

> If the conditions, materials and/or forces A, B, C, and D come together in such-and-such a way, then the outcome will be 6.7294874X. If variables P, Q, and R are substituted for A, C, and D, then the outcome will be 2.1 milligrams of tetrahydrocannabinol in combustion.

Hyde then applies this simple example of the predictive power of scientific theory to the highly assured predictions of climate warming by supposedly infallible science and scientists and their unreliable computer models.

In 1999, these scientists said that warming would wipe out the Great Barrier Reef. In 2000, they said that Britain would no longer see snow during winter. In 2001, they predicted starvation from failing grain crops in India. From 2003 to 2005, they concluded that the drought then occurring in Australia would be permanent and Sydney dwellers would have nothing to drink. In 2006, they predicted unprecedented severe cyclones and hurricanes. In 2008, they said that by 2013, there would be no more Arctic ice cap, that we would be swimming with the otters at the North Pole. Hyde, along with the rest of the world, can boldly announce that none of these predictions came to pass. The reef is still there, as is the Arctic ice. Children make more snowmen than ever in Britain, and the rains returned to Australia with a vengeance. Thanks to the instantaneous and ubiquitous communications made possible by our smartphones and social networking, there is much greater awareness of the severe weather events that do occur than there was before, but in absolute terms, such events are neither more frequent nor more severe than they have always been.

Hyde then reminds us, "The mother of all predictions, that global warming was inexorable, had been debunked as of 2015 by the past seventeen years of the actual 'pause' in temperature measurements."

Even with this empirical evidence refuting climate alarmist claims, Hyde warns that this won't stop the persistence of politicians with a vested interest in insisting that climate change is a greater threat to humanity than ISIS, Iran, North Korea, unemployment, burning American cities, and negative economic growth combined, and he suggests that something is wrong at a deeper level with the way we are practicing and discussing science.

Hyde argues,

> Scientists, strictly defined, should have no agenda whatsoever other than the discovery of truth; truth of which no human being is the ultimate arbiter, but only Nature. ... Scientists who have come to believe that a certain theory is closer to the truth than any known alternatives have the right, indeed the duty, to defend that theory against any and all challenges. But the true scientist must always, without exception, maintain intellectual honesty and be prepared to abandon a theory if its predictive power cannot explain empirical data that does not fit and/or when a rival theory that seems to do a better job of explaining the subject phenomena (often in simpler terms) arises. Skepticism and openness to change and to challenge is the fundamentalist creed of the true scientist.

A theory that does not contain the terms of its own falsification is not a valid theory.

To provide known examples of how the science process should work, Jerome J. Schmitt ("Galileo Denied Consensus," Schmitt 2007b) starts by explaining the twenty-first-century nonscience methods practiced by climate alarmists to enforce consensus about global warming by branding deniers of the impending apocalypse as heretics who lack blind faith in the theology of the infallibility of climate computer models. To enforce the consensus, deniers are being threatened with loss of their positions, credentials, and titles. Foisting theories upon scientists and the public by means of verbal persuasion, elections, court orders, or intimidation is the opposite of the scientific method of determining the truth.

By contrast, Schmitt provides two premier revolutions in science to illustrate the role of experimentation in proving new theories. First, Einstein's theory of relativity is so complex and counterintuitive that few could fully understand it at the time of its publication, and that remains the case for the most part even today. It overturned the prior consensus view of the universe based solely on Euclidean geometry and Newtonian physics. While controversial at the time, it was soon accepted because the theory yielded testable predictions undertaken by clever experimentalists. This account (NASA 2004) explains how the theory was first tested.

> The central premise of Einstein's general theory of relativity is that all matter and energy moving through the universe are affected by curved space-time. This includes the path of light rays as they emerge from distant stars and make their way across the universe to our Earth-based telescopes and eyes. When their light passes near a massive body, such as a galaxy or our Sun, its path is deflected slightly

> In 1919, merely three years after Einstein published his theory, Frank Dyson (Great Britain's Astronomer Royal

at that time), Charles Davidson, and Arthur Eddington took on the challenge of observing and measuring this phenomenon. They compared photographs of a selected area of the night sky with photographs taken of the same area during a solar eclipse. Looking at these photographs, it became apparent that stars that should have been behind the Sun were actually visible during the eclipse. Their light was bending through curved spacetime around the Sun's mass. The result was limited by the short amount of time available to make the measurement during an eclipse (about four minutes), but it confirmed Einstein's prediction to within about 20 percent.

According to Schmitt, these measurements were taken during an expedition organized for the purpose to West Africa where the eclipse could be best observed. The expedition's progress was followed closely in the press. Einstein was met with instant and widespread professional and public acclaim upon this empirical confirmation of his theory.

Second, Darwin's theory of evolution illustrates well the need for reliance on experimental confirmation. Darwin's sweeping theory was revolutionary when propounded, and it is controversial even today. Darwin sought to overturn the consensus view of the origin of species derived from literal interpretations of biblical texts. The part of his theory describing the basic evolutionary mechanism (microevolution) is almost universally accepted today even among advocates of intelligent design theory (Staff 2007) precisely because it has been tested experimentally in many ways. For instance (Darwin 1869),

An example of evolution resulting from natural selection was discovered among "peppered" moths living near English industrial cities. These insects have varieties that vary in wing and body coloration from light to dark. During the 19th century, sooty smoke from coal burning furnaces killed the lichen on trees and darkened the bark. When moths landed on these trees and other blackened surfaces, the dark colored ones were harder to spot by birds who ate them and, subsequently, they more often lived long enough to reproduce. Over generations, the environment continued to favor darker moths. As a result, they progressively became more common.

Microevolution can be directly observed in laboratory experiments with fruit flies, mice, and bacteria because they have very short lifetimes. Darwin himself made a testable prediction that was vindicated only after his death. His insights into coevolution allowed him to foretell the discovery of a new species. In a famous example, he described an orchid from Madagascar that had a foot-deep nectar well that kept the sweet liquid far out of reach of all known butterflies and moths. But the existence of the flower led him to predict the existence of a specialized moth with a foot-long proboscis that like a straw could reach the deep reward. Indeed, after Darwin's death, researchers discovered just such an insect and named it the *Predicta* moth in honor of Darwin's educated guess.

Schmidt notes that unlike microevolution, macroevolution has long been a consensus view of most scientists and indeed most highly educated people. This aspect of the theory holds that life itself evolved randomly from a primordial soup and that all species evolved from this primordial soup. However, Schmitt cautions that in recent years, critics of macroevolution including some highly credentialed scientists have criticized Darwin's theory. The lack of experimental evidence for the production of entirely new species has been one of the principal grounds. Schmitt argues,

> The degree of success of two revolutionary scientific theories was predicated on confirmation by test results. All of science is advanced by the toil of experimenters

who cleverly design methods and apparatus (such as Galileo's novel telescope) to measure natural phenomena. It is solely by convincing and reproducible empirical results that science advances; there is no other method. Consensus has nothing to do with it.

Schmitt holds that scientific professionals who propose a theory recognize their own need for experiments to help affirm, refine, and advance it. While many physicists are intrigued by the tenets of string theory, they acknowledge that consensus has nothing to do with its validity.

In applying these basic concepts to the theory of global climate warming, Schmitt argues,

It is incumbent on climate scientists to make predictions derived from their theories about phenomena that can be observed over the next few years. Indeed, it is *their* scientific duty as proponents of the theory to provide the experimental evidence, and not the responsibility of the skeptics to disprove it. Those are the established rules of science.

Besides being able to conduct experiments on a particular theory to prove or disprove its assertions, the process of peer review is the cornerstone or holy grail of scientific integrity, the guarantee of quality in a scientific research paper. In "The Peer Review Problem" (Sheahen 2016), Thomas P. Sheahen argues that when a researcher accomplishes something, the concluding step is to publish the results in scientific literature.

Sheahen describes the peer-review process as beginning when the author of a scientific paper sends it to a journal for consideration. The editor of this journal then sends the paper to other scientists who are the peers (equally capable scientists) of the author and active in the same research area. The readers send to the editor their reviews with comments and suggestions for improvement of the paper. The reviewers remain anonymous relative to the author to avoid softening of criticism that might be due. Based on that input, the editor accepts the paper for publication, sends it back to the author for revision, or rejects it.

Sheahen acknowledges that most professional scientists have experienced all three outcomes at various times. By far, the most common occurrence is that the author is asked to clarify some technical points and consider some overlooked paper in the literature that has bearing on the topic. After the author revises the text to conform to the criticism of the reviewers, the paper customarily goes forward for publication.

Because of this process, Sheahen acknowledges that another scientist reading the journal has confidence that the research papers published in the journal are reliable and that other researchers can use the information to influence the direction of their own research. The -peer-review process is the best way known to scientists to advance their professional disciplines. Without the checks and balances of such a process, the scientific literature would deteriorate into just another news outlet, a cluster of press releases and advertisements.

Sheahen notes that,

> most academics are evaluated on the basis of their published papers and often, a major criterion for being promoted is a count of the number of peer-reviewed papers a professor had. That translates into prestige and salary, and hence, there is built-in pressure upon a junior faculty member to generate many papers. The slogan "publish or perish" is well known in academia. This practice holds across all academic disciplines including history, psychology, social sciences, and literature, not just sciences likechemistry, biology, or physics.

Sheahen notes that some journals are more prestigious than others and that there is strong competition to have a paper published in a more prestigious journal rather than in one that is considered less prestigious. For example, the hardest journal to get a physics paper accepted by is *Physical Review Letters,* and a professor who gets even one paper published there during his or her whole career is very proud of it. Biochemistry, medicine, geology, history, and so on each has a hierarchy of journals. For instance, the *New England Journal of Medicine* is America's most prestigious medical journal.

Sheahen cautions that for the peer-review system to work effectively, everybody in the loop has to be scrupulously honest. The authors must have reached their conclusions honestly and based on valid data; they have to strive to present new information accurately to readers. The editors must select unbiased reviewers to conduct the peer review. The reviewers must be objective and guard against letting their own biases creep into their evaluations. When all this happens, a paper eventually gets published that adds an incremental advance to the field.

A skeptical Sheahen then reminds his readers that human beings are imperfect and hence, the peer-review process doesn't always work ideally. The very essential elements of integrity and trust can easily be lost at several stages. Researchers are sometimes blind to their own mistakes, and peer reviewers can be variously misinformed, hostile, or motivated by hidden agendas. Editors might have their own narrow views about what they want in their journals, or they can select certain reviewers to obtain negative or positive evaluations. When things go wrong, bad information gets unduly promoted. Unfortunately, Sheahen acknowledges that the history of peer review is not unblemished.

An example of a discipline in which this peer-review process in not working is the evidence of corruption in the basic temperature records maintained by key scientific advocates of the theory of man-made global warming. In "ClimateGate: The Fix is In" (Tracinski 2009), Robert Tracinski discusses evidence that global warming skeptics had unearthed evidence (McIntyre 2009) that scientists at the Hadley Climatic Research Unit (CRU) at Britain's University of East Anglia had cherry-picked data to manufacture a hockey-stick graph showing a dramatic but illusory runaway warming trend in the late twentieth century.

Tracinski further discusses the Climategate scandal in which hacked CRU emails exposed not just the good climate research but also the nefarious activities going on at the CRU involving numerous leading British and American climate scientists outside of the CRU.

Tracinski explains that these emails show among many other things private admissions of doubt or scientific weakness in the global warming theory and acknowledgments that global temperatures had actually declined for the previous decade. One scientist asked, "Where the heck is global warming? ... The fact is that we can't account for the lack of warming at the moment and it is a travesty that we can't." Tracinski cites the fact they still couldn't account for it.

In "Climatologists Baffled by Global Warming Time-Out" by Gerald Traufetter (Traufetter 2009), a disgusted Tracinski questions where these people got their scientific education since where he comes from, if your theory can't predict or explain the observed facts, it's wrong.

Although Tracinski touches on other scandalous activity revealed by these hacked emails, what stood out most for him was the extensive evidence of the hijacking of the peer-review process to enforce global warming dogma. The goal of the peer-review process is to weed out research with obvious flaws or weak arguments that would simply reinforce groupthink and create a mechanism for an entrenched establishment to exclude legitimate challenges by simply refusing to give critics a hearing. This is precisely what has happened with climate change science research.

Since the very essential elements of integrity and trust can be lost at several stages in the peer-review process and since researchers are sometimes blind to their own mistakes, Rick Rinehart ("Scientists Behaving Badly," Rinehart 2012) cites a huge scandal (Tracinski 2012) that erupted in the global warming community when a scientist who was fancied an expert on ethical behavior confessed to identity fraud aimed at illegally obtaining documents to discredit the Heartland

Institute, a leading organization questioning the climate alarmists' gospel ("Peter Gleick Admits to Deception in Obtaining Heartland Files," Revkin 2012).

Rinehart cites this example as a lead-in to his discussion on what integrity in the science community should look like. He recounted his time in the late 1970s and 1980s when he worked at the University Press of Colorado (then known as the clunky-sounding Colorado Associated University Press). It focused on the Boulder academic community's strengths—physical and natural sciences.

> His director was quick to form personal and professional bonds with some of the leading scientists at the National Center for Atmospheric Research (NCAR), so Rinehart's group was able to publish seminal works on everything from hailstorms to the physical output of the sun.

> Led by Walter Orr Roberts and Jack Eddy, a member of his pantheon affirmed a link between solar variability and climate. Since he was first and foremost a solar physicist, his primary interest was the behavior of the sun in recent history, and by taking a unique interdisciplinary approach studying a gamut of artifacts from tree rings to the observations of Galileo, he was able to correlate a cold period such as the Little Ice Age with relatively quiet solar activity. Moreover, he debunked the notion that the sun's output was constant. As he concluded somewhat prophetically,

> It was one more defeat in our long and losing battle to keep the Sun perfect, or, if not perfect, constant, and if inconstant, regular. Why we think the Sun should be any of these when other stars are not is more a question for social than for physical science.

Rinehart believed that Eddy represented an NCAR that was very much about the sun and its influence on weather and climate, and whenever Rinehart called on him and others in the I. M. Pei–designed Mesa Laboratory, he felt completely confident that their scientific research was not being compromised by politics. Though many at NCAR undoubtedly leaned liberal (this is Boulder, after all), the integrity of science was paramount.

Rinehart is taken aback by how things have changed over the last thirty years not just at NCAR but also throughout the climate science community. He cannot image a Jack Eddy or a Walt Roberts ever fudging data to arrive at a preconceived conclusion. They would be thrilled to publish valid dissenting views just to keep the climate science pot boiling. In fact, Rinehart notes that Eddy's landmark study of solar variability was so controversial that it made the cover of *Science* in 1976. Imagine the present-day stewards of the magazine giving the cover to the likes of Roy Spencer (Spencer 2018a).

Rinehart identifies a scientist who reflects the honor of Walter Orr Roberts and Jack Eddy, Judith Curry (Curry 2018) at the Georgia Institute of Technology. Not sold on either the skeptics' position or that of the IPCC, she is the adult in the room when the food fights start to break out in the climate kindergarten. Her response to the Gleick ("Peter Gleick Admits to Deception in Obtaining Heartland Files," Revkin 2012) confession was to write a thoughtful blog (Curry 2012) on the meaning of integrity though her disappointment with Gleick is palpable. She writes, "Gleick's 'integrity' seems to have nothing to do with scientific integrity, but rather loyalty to and consistency with what I called the UNFCCC/IPCC ideology."

Rinehart understands that students at Georgia Tech can take courses from Curry and others that present divergent points of view on the current climate controversy and in an exercise in critical thinking decide for themselves. How wonderfully sane and dignified.

14

A Historical Version of Earth's Climate

IN "CAN we make better graphs of global temperature history?" (Schmidt 2014) that Gavin Schmidt posted on *Real Climate, Climate Science from Climate Scientists*, March 13, 2014, he expresses frustration with and seeks support from the scientific community to

> collectively produce some coherent, properly referenced, open-source, scalable graphics of global temperature history that will be accessible and clear enough that can effectively out-compete the myriad inaccurate and misleading pictures that continually do the rounds on social media.

The current problem with the whole climate discussion is that if the self-identified climate scientists cannot agree on the history of geological climate history, how in the world can they understand the current climate conditions? More important, how can they establish the predictability of future drivers of climate conditions?

Not so long ago, the scientific approach to getting agreement was through peer review, where the known historical data sets are available in a public forum so open research can take place on how each of these data sets can be amalgamated for use in scientific climate research. This varied research needs to take place in an environment in which researchers are free from ad hominem attacks by agenda-driven operatives. In the past, science was about discovering the truth, not about taking sides, with its final form evolving to the needs of the scientific community.

In the absence of established and agreed-upon climate history, the following provides a version of climate history and methods available.

As described in Wikipedia,

> Temperature reconstructions based on oxygen and silicon isotopes from rock samples have predicted much hotter Precambrian sea temperatures. These predictions suggest ocean temperatures of 55–85 °C during the period of 2,000 to 3,500 million years ago, followed by cooling to more mild temperatures of between 10-40 °C by 1,000 million years ago. Reconstructed proteins (Wikipedia, ASR) from Precambrian organisms has also provided evidence that the ancient world was much warmer than today.

> However, other evidence suggests that the period of 2,000 to 3,000 million years ago was very generally colder and more glaciated than the last 500 million years. This is thought to be the result of solar (Wikipedia, Sun) radiation approximately 20 percent lower than today. Solar luminosity (Wikipedia, Solar) was 30 percent dimmer when the Earth formed 4.5 billion years ago, and it is expected to increase in luminosity approximately 10 percent per billion years into the future.

> On very long-time scale, the evolution of the sun is also an important factor in determining Earth's climate. According to standard solar theories, the sun will

gradually have increased in brightness as a natural part of its evolution after having started with an intensity approximately 70 percent of its modern value. The initially low solar radiation, if combined with modern values of greenhouse gases, would not have been sufficient to allow for liquid oceans on the surface of the Earth. However, evidence of liquid water at the surface has been demonstrated as far back as 4,400 million years ago. This is known as the faint young sun paradox (Wikipedia, FYSP) and is usually explained by invoking much larger greenhouse gas concentrations in Earth's early history, though such proposals are poorly constrained by existing experimental evidence.

The Neoproterozoic era (1,000 to 541 million years ago) (Wikipedia, Neo), provides evidence of at least two and possibly more major glaciations. The more recent of these ice ages, encompassing the Marinoan & Varangian glacial maxima (about 560 to 650 million years ago, Wikipedia, Snowball), has been proposed as a snowball Earth (event with continuous sea ice reaching nearly to the equator. This is significantly more severe than the ice age during the Phanerozoic (Wikipedia, Phanerozoic). Because this ice age terminated only slightly before the rapid diversification of life during the Cambrian explosion (Wikipedia, Cambrian), it has been proposed that this ice age (or at least its end) created conditions favorable to evolution. The earlier Sturtian glacial maxima (~730 million years) (Wikipedia, Sturtian) may also have been a snowball Earth event, though this is unproven.

The changes that lead to the initiation of snowball Earth events are not well known, but it has been argued that they necessarily lead to their own end. The widespread sea ice prevents the deposition of fresh carbonates in ocean sediment. Since such carbonates are part of the natural process for recycling carbon dioxide, short-circuiting this process allows carbon dioxide to accumulate in the atmosphere.

The Phanerozoic eon (Wikipedia, Phanerozoic), encompassing the last 542 million years and almost the entire time since the origination of complex multi-cellular life, has more generally been a period of fluctuating temperature between ice ages, such as the current age, and "climate optima," similar to what occurred in the Cretaceous (145 to 66 million years ago)(Wikipedia. Cretaceous). Roughly 4 such cycles have occurred during this time with an approximate 140 million year separation between climate optima. In addition to the present, ice ages have occurred during the Permian-Carboniferous (359 to 299 million years ago)(Wikipedia, Carboniferous) interval and the late Ordovician (485 to 444 million years ago) (Wikipedia, Ordovician) to early Silurian (444 to 419 million years ago)(Wikipedia, Silurian). There is also a "cooler" interval during the Jurassic (201 to 145 million years ago)(Wikipedia, Jurassic) and early Cretaceous, with evidence of increased sea ice, but the lack of continents at either pole during this interval prevented the formation of continental ice sheets and consequently this is usually not regarded as a full-fledged ice age. In between these cold period, warmer conditions were present and often referred to as climate optima. However, it has been difficult to determine whether these warmer intervals were actually hotter or colder than occurred during the Cretaceous optima.

During the later portion of the Cretaceous (Wikipedia, Cretaceous), from 100 to 66 million years ago, average global temperatures reached their highest level during the last 200 million or so years. This is likely the result of a favorable configuration of the continents during this period that allowed for improved circulation in the oceans and discouraged the formation of large-scale ice sheets. Perhaps the visible anecdotal evidence of high temperatures during this period was the occurrence of deciduous forests extending all the way to the poles (Wikipedia, PFOTC).

> In the earliest part of the Eocene (Wikipedia, Eocene) period (56 to 33.59 million years ago), a series of abrupt thermal spikes has been observed lasting no more than a few hundred thousand years. The most pronounced of these, the Paleocene-Eocene Thermal Maximum (Wikipedia, PETM) is visible ... These are usually interpreted as caused by abrupt releases of methane from clathrates (frozen methane ices that accumulate at the bottom of the ocean), though some scientists dispute that methane would be sufficient to cause the observed changes. During these events, temperatures in the Arctic Ocean may have reached levels more typically associated with modern temperate (i.e. mid-latitude) oceans.

During the PETM, the global mean temperature seems to have risen by as much as 5 to 8 degrees C (9 to 14 degrees F) to an average temperature as high as 23 degrees C (73 degrees F) in contrast to the global average temperature of today at just under 15 degrees C (60 degrees F). Geologists and paleontologists think that during much of the Paleocene and early Eocene eras, the poles were free of ice caps, and palm trees and crocodiles lived above the Arctic Circle while much of the continental United States had a subtropical environment.

The reconstruction of the past 5 million years of climate history based on oxygen isotope fractionation in deep-sea sediment cores (serving as a proxy for the total global mass of glacial ice sheets) fitted to a model of orbital forcing (Wikipedia, OF) and to the temperature scale derived from Vostok (Wikipedia, Vostok) ice cores following Petit et al. (1999).

The last 3 million years (Wikipedia, Pleistocene) have been characterized by cycles of glacials and interglacials within a gradually deepening ice age. Currently (Wikipedia, Holocene), the earth is in an interglacial period beginning about 20,000 years ago (Wikipedia, LGM).

> The cycles of glaciation involve the growth and retreat of continental ice sheets in the Northern Hemisphere and involve fluctuations on a number of time scales, notably on the 21 kiloyear (thousand), 41 kiloyear and 100 kiloyear scales. Such cycles are usually interpreted as being driven by (Wikipedia, OF) predictable changes in the Earth orbit known as Milankovitch cycles (Wikipedia, MC). At the beginning of the Middle Pleistocene (Wikipedia, MP) (800,000 years ago, close to the Brunhes–Matuyama geomagnetic reversal Wikipedia, MGR) there has been a largely unexplained (Wikipedia, 100,000) switch in the dominant periodicity of glaciations from the 41 kiloyear to the 100 kiloyear cycle.

> The gradual intensification of this ice age over the last 3 million years has been associated with declining concentrations of the greenhouse gas carbon dioxide. Decreased temperatures can cause a decrease in carbon dioxide. As explained by Henry's Law (Wikipedia, HL), carbon dioxide is more soluble in colder waters, which may account for 30ppmv of the 100ppmv decrease in

carbon dioxide concentration during the last glacial maximum.

Similarly, the initiation of this deepening phase also corresponds roughly to the closure of the Isthmus of Panama by the action of plate tectonics. This prevented direct ocean flow between the Pacific and Atlantic, which would have had significant effects on ocean circulation and the distribution of heat.

In the research paper "Geologic Global Climate Change" (Nahle 2007), Nasif Nahle tackles the carbon dioxide climate forcing issues in more detail. He argues that scientific studies have shown that atmospheric carbon dioxide in past eras reached concentrations that were twenty times higher than the current concentration. At Biology Cabinet (Lederman 2001), scientists maintain that the changes we have observed since 1985 have been natural and that human beings cannot delay or stop the advance of these changes but can only adapt to them. In addition, the group has shown that the changes being observed are the result of natural cycles (Nahle 2007) that have occurred many times before.

Nahle's research show that in prehistoric times, during the Permian in the Paleozoic eras for example, the concentration of carbon dioxide dropped below 210 ppmV. Throughout the Permian period, plant and animal species diverged and diversified as never before. Dinosaurs prospered and predominated all other orders of vertebrates. Coniferous plants first appeared in the Permian. The change of atmospheric temperature at the time of the Permian was around 10 degrees C. By comparison, the current change of global temperature is only 0.52 degrees C while the concentration of atmospheric carbon dioxide is around 400 ppmV.

If the global temperature is dependent on carbon dioxide, the change of temperature at present would be around 10 degrees C or higher as it was during the Permian Period.From the early Triassic to the middle Cretaceous, the concentration of atmospheric carbon dioxide was similar to its current density. From the late cretaceous to the early Miocene, the concentration climbed above 210 ppmV. During the Holocene period, the concentration has oscillated from 210 ppmV to 385 ppmV."

Nahle speculates that the concentration of atmospheric carbon dioxide will increase normally in the course of the next 50 million years to 1050 ppmV or 2500 ppmV.

Scientists have also observed that the concentration of atmospheric carbon dioxide increases several centuries after glaciations. Perhaps this is due to the fact that most plants perish at subzero temperatures and that plants are organisms that capture carbon dioxide from their surroundings to make food.

Scientists have observed that the concentration of atmospheric carbon dioxide increases during periods of warming. However, an increase in temperature always precedes an increase in carbon dioxide, which generally occurs decades or centuries after any change of temperature.

Nahle has not observed,

an increase in the concentration of carbon dioxide to have preceded a period of warming. This latter phenomenon occurs because when oceans absorb more heat from an increase in the amount of direct solar irradiance incident upon the earth's surface, they release more carbon dioxide molecules into the atmosphere. Nevertheless, most drastic increases in carbon dioxide concentration occur decades or centuries after the oceans have warmed up. For example, the present increase of atmospheric carbon dioxide was caused by an extraordinary increase in solar activity in 1998, which warmed up the El Niño South Atlantic Oceanic Oscillation.

These increases in concentration of atmospheric carbon dioxide offer optimal conditions for the development and evolution of living beings on Earth. Human beings should adapt to these natural changes by means of science and technology.

As for continental flooding area on the geologic timescale, Nahle's research concludes that, higher sea levels generally correspond with periods of warming while lower sea levels marry generally with periods of cooling. The lower sea levels are explained by a reduction of sea water as the oceans ice up at the poles, although the group has noticed also that the sea level response is sometimes negative with respect to warming or cooling of the atmosphere.

Further, on the topic of concentration of atmospheric carbon dioxide and change of sea levels, Nahle's research shows that,

> increases of the concentration of atmospheric carbon dioxide always follow drops in sea level. Since the drops in sea level are caused by oceans cooling, the load of carbon dioxide released to the atmosphere is much smaller than the load released to the atmosphere when oceans are warming.

Something else worth considering in this comparison by Nahle is that,

> the concentration of atmospheric carbon dioxide has decreased as the oceans have cooled over geological time. The correlation between both phenomena—decrease of the concentration of atmospheric carbon dioxide and lowering of the sea level—is supporting evidence that the oceans are the secondary driver of the earth's climate.

For Nahle, "the Sun is the primary driver of climate on our planet." Nahle concludes from his research that,

> the evidence points to a natural climate change states which happens sequentially in two main climate periods, icehouse and warmhouse.

To provide further details, Nahle identifies and names the succession of four natural climate phases known as *Transgression, Highstand, Regression* and *Lowstand*.

> The *Transgression* phase consists of a rising Sea Level, flooding continental areas. *Highstand* is a phase where the marine level remains relatively stable but oscillating into the *Transgression* phase. The *Regression* phase consists of a gradual diminution of the marine level, leaving a greater area of the continents uncovered. The phase of *Lowstand* consists of a permanence of low marine level. Currently, the earth is passing through a *Lowstand* phase, which will revert to *Transgression* phase. The succession of these phases show the earth is cooling.
>
> At the moment, the area of continental flood is almost 7%; according to climatic succession,

Nahle expects,

> the area of continental flood to increase to almost 10%,
> but never so massive that it will put human populations
> in danger, as the IPCC has taken to suggesting almost
> every day.

15

Arguments against Man-Made Climate Change

IN "TRUMP, the Times, and the Coming Eco-Apocalypse" (Cashill 2016), Jack Cashill cites an interview with Trump in November 2016 by Thomas Friedman with the *New York Times* wanting to sound the alarm on carbon dioxide causing climate warming turning Trump's "oceanside [golf] courses into ocean-floor courses." Jokingly, Trump suggested a rise in sea levels just might increase the value of his Doral golf course given that it is about ten miles inland.

For Friedman, this was no laughing matter. "It's really important to me," he huffed. Said Trump, "I'm looking at it very closely, Tom. I'll tell you what. I have an open mind to it." This was hardly a flip-flop. Trump touched on the Climategate scandal and told the *Times* crew something no one with power likely ever told them before: "A lot of smart people disagree with you."

Cashill noted an unsatisfied publisher Arthur Sulzberger ratcheting up the hysteria. "We're living on an island, sir," he warned Trump. Sulzberger then shared with Trump a related fear, that of storms. Trump tried to comfort him. "We've had storms always, Arthur," he replied. But Sulzberger refused to hear it. "Not like this," he said.

Cashill suggested that although Trump did not question Sulzberger's definition of "like this," he could have safely done so. Florida had gone a record eleven years without a single hurricane strike before the modest category 1 Matthew struck northern Florida in October 2016.

Similarly noted Cashill, according to the Weather Channel, the even more modest Hermine "eclipsed the longest drought on record, dating to 1886," when it entered the Gulf of Mexico in September 2016. Before Hermine, not a single hurricane had entered or developed in the gulf since 2013. Apparently, with the wiggle room that the phrase "climate change" allows, "like this" can mean either too many storms or too few.

Cashill argues that on the subject of sea levels and shrinking beaches, beaches constantly change as does the climate; they are subject to tides, storms, and erosion. It is reasonable to assume that some are a bit smaller than they once were; some are larger. Unlike the climate, however, one can photograph a beach, and people have been doing so routinely for more than a century. If there were a case to be made for rising sea levels, the *Times* could better convince Trump and the rest of the population through hard photographic evidence than through questionable projections in their publications and by others.

For Cashill, however, the *Times* people have a faith that transcends evidence. They will surely cling to their core beliefs as gospel even when confronted with the obvious. "Facts aren't necessary," said late author Michael Crichton on the subject of environmental doomsayers. "It's about whether you are going to be a sinner, or saved. Whether you are going to be one of the people on the side of salvation, or on the side of doom. Whether you are going to be one of us, or one of them." (Crichton 2003) As the *Times* people are finding out, if there ever were a "one of them," it is surely Trump.

And for the last few decades, political leaders and not just Trump have been arguing about the science of carbon dioxide and how much to reduce carbon dioxide emissions if at all to avert the consequences of man-made global warming as advocated by the climate alarmists. But more and more, science is showing that the earth's climate is dominated by natural climate cycles driven mainly by solar activity.

Man-made carbon dioxide emissions play only a small role in these cycles.

In "Who's afraid of Global Warming?" (Dunn 2007), J. R. Dunn argues that this is not the first time climate warming has occurred on Earth and that it can be counted on as a recurring phenomenon. Dunn cites one such episode that took place in relatively recent historic time, the LCO, from about 950 to 1250. During the LCO, worldwide temperatures rose by 1 to 3 degrees C with records from the era being abundant and easily available.

Dunn then addresses the major dire climate events being advocated by these alarmist groups. The most popular topic put forward by climate alarmists is sea level rise with spectacular visuals of an underwater New York City as predicted in Gore's *An Inconvenient Truth* or from a meter in the 2000 IPCC report to forty feet as predicted by Australian activist Tim Flannery.

And during the LCO, which lasted over three centuries, the highest oceanic level was eighteen inches above the previous norm. According to Dunn, that "foot-and-a-half may sound like quite a lot, but the damage it caused appears to be minimal. There are no records of massive flooding either in Europe or elsewhere. No seacoast villages were relocated that we know of."

To check on beach erosion during the LCO period, Dunn cites the *furdustrand*, "wonder strand," so named by the Vikings who were the first Europeans to come across it about AD 1000. The furdustrand or now known as Wonderstrands, is a white sand beach close to forty miles long and in places two hundred feet wide. It is spectacular, and it would be a lot more widely known if it was anywhere on earth more accessible than Labrador, Canada.

Dunn's points out that the rising sea levels of the LCO, the retreating levels of the Little Ice Age, and the return to higher levels since 1850 appear not to have harmed it one bit. With these two examples and others not cited here, Dunn concludes that fears of disappearing beaches can be dismissed.

The next fear being promoted by the climate alarmists is that coral reefs will be wiped out by global warming. Dunn has difficulty getting his head around this claim since coral reefs thrive in the warm waters of the tropics. The best known is the Great Barrier Reef off Australia's tropical northeastern coast, and of course, the entire South Pacific is dotted with atolls that began their careers as exactly such reefs. Dunn challenges anyone to identify any islands in the cold waters of the Arctic or the Pacific Ocean that can empirically demonstrate rising sea waters affecting their shoreline.

Dunn notes that the same coral reefs existed during the LCO and appear not to have been affected by the large-scale warming that occurred at the time. To support this claim, Dunn notes that there are no beds of dead ancient coral visible and no legends of mass die-offs by Melanesians or other native peoples (dying coral would have deprived fish of a safe environment leading to a drop in the food supply). Dunn concludes that no such event happened.

In fact, recent research has clearly demonstrated that sewage runoff is the actual culprit that poisons reefs off both Australia and the United States. Runoff of fertilizer, pesticides, and other chemicals may also have an adverse effect.

Dunn then turns to climate alarmists claiming earth's warming is resulting in mass extinction of living species. These alarmists are making the claim that up to a quarter or a half of all species are disappearing, though there's no concrete evidence of a single species actually being threatened by warmer temperatures. Dunn concludes that as with much warming hysterics, this seems to be sheer speculation based on the premise that certain niche organisms will die out as their marginal environments are changed.

Dunn argues the problem with this thesis is that no species appears to have vanished as a result of the LCO. Even though warming and cooling has occurred continually throughout the geological history of the planet earth, it's safe to assume that most organisms have developed means of dealing with these slow environmental changes.

Severe storms are mentioned for *pro forma* reasons as much as anything. We're all aware (much as the liberal big media has chosen to neglect the fact) that the 2006 fall hurricane season,

> predicted to be second only to the Day of Wrath in violence, was a complete washout with not a single serious hurricane troubling American shores. This was a grave disappointment to the greens after 2005's wild roller-coaster ride.

Dunn then changes focuses to another consequence from global warming, increasing severity of storms, particularly tornados and hurricanes. The alarmists' severe storm thesis is a result of their junior high science: the atmosphere is a heat engine, so if you add more heat, there will be more activity with storms growing in frequency, duration, and violence with no perceptible upper limit... In truth, most warming occurs at higher latitudes effectively erasing differences in atmospheric temperature and meliorating weather.

Dunn again references the LCO period, which demonstrated periods of generally calm and predictable weather with lengthy summers, gentle winters, and fierce storms relatively rare and all the more striking for that.

> This calm literally lasted for centuries and enabled the Vikings to carry out their explorations in open boats at very high latitudes, areas afflicted with horrible weather even to this day. On the other hand, numerous violent storms reappeared when the climate cooled in the late thirteenth century with terrifying results. Consider the fate of Winchelsea, an English port swallowed by the waves of the English Channel during a days-long rainstorm in 1297. Even worse, there were crop failures caused by dismal weather

all across Europe that resulted in repeated general famines.

Dunn argues that it is a cooling climate that leads to foul weather.

But it is the alarmists' claims of the melting of the world's major ice sheets—those of Greenland and Antarctica—that is nearly pure fantasy. Dunn calculates it would take a millennium of continuous hot weather to make a dent in either. The LCO, which lasted a little over three centuries, failed to leave much of a mark.

Dunn argues that it is possible that,

> warming may actually increase the thickness of the continental ice sheets by increasing evaporation, which then falls as snow. Meteorological observations seem to indicate that this is happening to both ice sheets and suggest this may be part of a planetary homeostatic system that keeps things in rough balance.

For Dunn, the WHO and its newfound mission to conflate climate warming with predictions will bring about a vast increase in disease. Dunn refers again to the LCO period. While by no means disease free, the Medieval Warm Period was as close to it as any era before the pre-modern world can show. The black plague,

> the chief dread of the period, completely retreated from Europe to its original home in central Asia (evidently, rodents in the Caucasus have adapted to the plague bacillus and serve as a steady, living reservoir). There are no outbreaks of plague on record during the LCO and few of other diseases. This was the direct result of a combination of gentle weather and good harvests; well-fed people tend to have robust immune systems.

Dunn further notes that movement of the Vikings and the Mongols, who burst into Europe before the end of the era,

> were not followed by massive exchanges of diseases, which normally occurs when cultural bubbles are broken after long periods of isolation. Consider the varied and deadly plagues that killed much of the native population of Mexico after the Spanish invasion.

Dunn further cautions that the modern era is different; cheap jet travel allows easy and quick transmission of diseases as was shown with the SARS outbreak in 2003, which leaped from China to Toronto in a matter of days.

It's difficult to discern the exact nature of the purported relationship between warming and economic performance, and the climate alarmists' rhetoric offers little assistance. Dunn guesses that the specter of a crashed economy is simply added on as a matter of course, as a kind of Fifth Horseman armed with pink slips and foreclosure notices rather than scythes and swords.

Dunn is certain there's nothing inherent in any warming scenario that would lead to the economy going south. It must be all those plagues and storms.

> As for economic performance during climate warming or lack thereof, Dunn suggests that it is not easy to compare a modern economy with that of the feudal epoch except to say that the LCO appears to have encompassed an era of general good fortune. A peasant culture requires little more than plentiful food and roofs that don't leak, and the LCO had both.

That ended when the cooling came at the close of the thirteenth century. The encroaching cold was accompanied by the medieval depression, which lasted for over two centuries. The trigger was declining harvests and the plagues that followed.

To provide a reasoned argument on the state of carbon dioxide being the main driver of climate warming, S. Fred Singer ("Climate Realism," Singer 2012b) cites an example of two essays that represents a civil peer-review interaction. One, by a group of sixteen prominent scientists ("No Need to Panic about Global Warming," Allegre et al. 2012) was published in the *Wall Street Journal* on January 27, 2012. The second was an essay by Yale professor William D. Nordhaus ("Why the Global Warming Skeptics are Wrong," Nordhaus 2012) in the *New York Review of Books* on February 22, 2012, which challenges the arguments put forward by the Allegre et al. group.

Singer makes clear before making his professional arguments related to this particular discussion that he considers himself a climate warming skeptic, not a climate warming denier—he is open to scientific evidence presented from either side and will weigh the facts against his own professional experience. With this clear position in mind, Singer acknowledges he has known Bill Nordhaus for about 40 years; he certainly is no wild-eyed climate alarmist, but rather a highly respected specialist in environmental economics. Through his association with the UN climate-science panel, he is familiar with the main arguments supporting the IPCC's contention that human activities, mainly rising CO_2 levels from energy generation, have been responsible for much of past warming. He does not question this IPCC claim; however, I have no reason to believe that he supports any of the drastic CO_2-mitigation schemes—be they carbon sequestration or alternative "green" energy projects—or that he has illusions about the efficacy of the Kyoto Protocol or similar measures of international control.

Singer made the following points in a letter he published in the same *New York Review of Books* on August 16, 2012, in which he simply tries to address questions Nordhaus posed in his *NYRB* essay published on February 22, 2012. As Singer stated in this article, Nordhaus "wanted my response to reach *NYRB* readers, typically liberal academics, lawyers, and teachers."

Singer begins his critique of the essays published by the two competing views of climate warming by "noting they may have over-looked two crucial points in their debate ... Evidence for anthropogenic global warming is problematic [and] a modest warming is likely to be beneficial—not damaging."

The following five questions where posed by Nordhaus with answers provided by Singer in his article.

1. Is the planet in fact warming?

This crucial question cannot be answered honestly unless one specifies the time interval referred to. Clearly, the climate has warmed since the last Ice Age. It has also warmed since about 1850 in recovering from the Little Ice Age (roughly 1400–1800). But it has not warmed since the Medieval Warm Period of 1,000 years ago or since the Holocene Optimum, which reached even higher temperatures 5,000–8,000 years ago. Nor has it warmed during the past decade.

Coming closer to the present, there is warming between 1910 and 1940, which is real but not caused by human activities. Most would agree that the earth's surface cooled slightly between 1940 and 1975 even though CO_2, a greenhouse gas, had been steadily increasing during that period. Temperature data show a sudden, unexplained jump around 1976/77. Surface weather stations then reported a modest increase in temperature up to the year 2000 though different analyses disagree on details and have been frequently revised.

Many people, including Nordhaus, tend to identify this reported increase as caused by the almost parallel increase in CO_2. In its summary, the 2007 IPCC report states explicitly that this reported (surface) warming trend is sure (>90 percent) evidence for AGW.

The atmosphere, both over land and ocean, did not warm during this same post-1978 period even though atmospheric theory and every climate model predicts that the tropical atmosphere should warm nearly twice as rapidly as the surface does. This atmospheric evidence comes from instruments in weather satellites producing the only truly global data and independent of thermometers in balloon-borne radiosondes.

In 2000, the NAS assembled a team of distinguished scientists to discuss the puzzle of surface warming in the absence of an atmospheric warming trend. However, their report, "Reconciling observations of global temperature change" (Science 2000) could not reconcile the disparity [observed versus predicted].

An analysis of ocean data has shown no significant warming during the period 1978–2000. Independent nonthermometer data (so-called proxies such as tree rings, ice cores, ocean sediments, stalagmites, etc.) also show no warming trend between 1978 and 2000. Significantly, there was no warming for this decade. All this in spite of constantly rising CO_2 levels.

The inescapable conclusion—or as Singer qualifies perhaps suspicion—is that land-based weather stations may have been reporting just local temperature increases but that there is negligible global warming. If correct, this surmise would remove the main evidence for the IPCC's claim about the existence of appreciable AGW.

2. Is CO_2 a pollutant?

Lawyers might say, "Yes, that is what the Supreme Court ruled in 2007," but scientists are not so sure. A pollutant by definition must produce harmful effects. CO_2 is a natural constituent of the atmosphere; it is nontoxic and invisible and has no physiological effects scientists know of even at high concentrations. Its definition as a pollutant relies entirely on its alleged causation of significant global warming and on the additional assumption that a warmer climate is damaging.

Singer further notes that CO_2 is Mother Nature's plant fertilizer. The world's important crop plants developed when CO_2 levels were much greater than today's CO_2 levels. Innumerable experiments have demonstrated that higher CO_2 concentrations are beneficial for plant growth and therefore benefit global agriculture; plants grow faster and require less water under such conditions. All of this is well known to agricultural experts and to the owners of commercial greenhouses, who often raise CO_2 levels artificially to increase productivity. Perhaps we should be grateful to China, the world's largest emitter of CO_2.

3. Are we seeing a regime of fear for skeptical climate scientists?

Singer considers himself fairly senior in his profession and is not much affected by the animosity toward skeptics revealed by the leaked emails from Climategate. However, Singer is suspicious of having lost friends in the academic community and has had considerable difficulty in getting technical papers published in journals whose editors have openly expressed their bias. Singer's real concern is for younger scientists who are just trying to establish their professional careers.

4. Are the views of mainstream climate scientists driven primarily by the desire for financial gain?

This is a leading question; Singer makes the assumption that scientific curiosity is the main driving force. Financial gain may be only one of several additional factors along with prestige and academic advancement, invitations to important conferences, prizes, etc. However, Singer points to the large sums, over $20 billion during the past decade, that the government has spent on climate research of which only a tiny fraction has gone to skeptics. Singer also notes the multimillion-dollar grants to mainstream climate scientists by private foundations and even by oil companies such as Exxon and BP. Not surprisingly, the number of scientific publications is roughly proportional to this level of financial support.

5. Is it true that more CO_2 and additional warming will be beneficial?

Singer's answer is yes. Nordhaus correctly states that net benefits should be maximized. This is mathematically equivalent to the well-known result that one should increase pollution control as long as marginal benefits exceed marginal costs. As an expert economist, however, Nordhaus should expand his discussion of more important points.

The discount rate plays a crucial role in the present case, where costs are incurred today, while benefits may be realized a hundred years hence. Nordhaus himself uses realistic discount rates of 4 percent, but he should be more critical of others such as Lord Nicholas Stern, who uses discount rates close to zero, which severely skews any cost-benefit analysis by greatly overestimating the present dollar value of benefits.

Further, one must ask if there is really any net damage at all from a warmer climate. Singer wonders why Nordhaus never mentions the work of Yale resource economist Robert Mendelsohn and his twenty-three economist colleagues whose acclaimed book concludes that a modest warming and higher CO_2 levels would actually enhance GDP and thus raise average income, prosperity, and general welfare.

For Singer, it should be obvious that if a warmer climate produces positive net benefits rather than damages, in principle, one cannot even conduct a cost-benefit analysis. Nor should one try to mitigate emissions of CO_2 in any way; our current policies are simply misguided.

In another scientific misdirect by the climate alarmists, James Lewis ("Is there an average global temperature?" Lewis 2007a) cites the article "Does a Global Temperature Exist?" (Essex et al. 2006) that appeared in *The Journal of Non-equilibrium Thermodynamics* and authored by Christopher Essex, Ross McKitrick, and Bjarne Andresen. They raise the point that there may be no global warming because there is no such thing as a global temperature. This controversial conclusion is based on the fact that "the earth atmosphere is not a homogeneous system"; that is to say, it is not a glass lab jar in a high school physics lab.

According to Bjarne Andresen et al.,

> It is impossible to talk about a single temperature for something as complicated as the climate of Earth. A temperature can be defined *only for a homogeneous system*. Furthermore, the climate is not governed by a single temperature. Rather, differences of temperatures drive the processes and create the storms, sea currents, thunder, etc. which make up the climate. (italics added)

Lewis notes that the cited journal itself deals with energy systems that are too complex to come to equilibrium unlike a cup of hot tea that behaves in a highly predictable way. A lot of important physical systems such as the climate appear to be nonequilibrium systems. They are not well understood, which is why they are a hot frontier topic in physics.

From this paper (Lewis 2007a), Lewis explains,

> Mathematically, there are several different "measures of central tendency," which is what an "average" really is. When scientists think about average global temperature, they are usually thinking about the arithmetic mean. But there is also a geometric mean, a mode, a median, and more complicated expressions that can be used as numerical indices for the heat content of a physical system. But as Andresen, et al. points out, which of those "averages" used *depends upon your model of the atmosphere.*

> The current evidence cited for global warming could even mean a decrease in the physical heat density of the atmosphere, if a different mathematical average is used. And because the climate is driven by *differences* in heat between different regions --- leading to the daily weather, as well as hurricanes and snow storms --- the right predictor for global climate may not be an average heat density at all, but rather the regional differences in heat content. Weather systems flow from high to low pressure regions, which are in turn dependent upon complex heat exchange mechanisms.

> All the standard arguments for global warming rely upon conventional "equilibrium" models of the atmosphere, all of which may be false.

Another pseudoscience presented as fact by the climate alarmists is rising sea levels. In "Measuring Sea Level Is a Suspect Art" (Hampson 2016), Spike Hampson states that a scientific discussion on rising sea levels can proceed only if the skeptics and alarmists "actually know what the current sea level is and how it has been behaving." Hampson further states that both sides tend to accept that data published by NASA and NOAA regarding measured year-to-year changes in sea level help support their respective positions on the topic.

Hampson is skeptical of these measurements published by NASA and NOAA; he argues that there are conceptual problems associated with identifying the sea level given the technological limitations on how these measurements are made.

Hampson notes that NOAA and NASA claim that the sea level is rising about an eighth of an inch per year or more or less equivalent to the stacked height of just two quarters. The two measurement systems used to arrive at this figure are tide gauges in coastal areas and satellite pings over the oceans. Hampson argues that neither method can record measurements of sea level any better than to a few inches of precision.

According to Hampson, NOAA and NASA argue away this problem of instrumental crudity by assuming that measurement imprecision is unbiased in its distribution and that a large number of measurements taken in a very short span of time can be averaged to yield an estimate of sea level that is more precise and more accurate than the actual data obtained from the individual instruments.

Hampson argues that there are several problems associated with measuring sea level. First, the sea never lies flat. Waves or swells constantly vary the height of the sea surface, and that variation typically exceeds a foot or two at least a handful of times every minute. Even when the sea appears to be calm, the subtle undulations of passing swells work the same magic. But usually, the sea is not calm, and the actual sea level height at any particular point is unpredictable.

Tides are marginally more predictable, but they change the level of the sea perhaps twice as much on average as do waves and swells. They do so at a slower pace alternating between peaks and valleys only once or twice a day, but when trying to ascertain the global average sea level for a year, this is still an enormous amount of noise to filter out of the computation.

As with calculating an average temperature for the earth, Hampson argues that the practice of averaging multiple sea level measurements is not appropriate when scientists use it for canceling out the effects of waves, swells, and tides. A computed average (mean) deserves credibility only if the data values are distributed normally (so as to describe the well-known bell-shaped curve). But waves, swells, and tides clearly do not vary sea level according to this limitation. Instead, they linger at the extremes (e.g., high and low tides) and quickly pass by the middle range, where the computed average is bound to fall.

Hampson argues that there are additional measurement problems when calculating changing sea level. There is no way to determine whether it is the same for different locations or if the earth's land masses are not stationary. Geologists believe that the earth's crust floats on a superheated liquefied rock known as the mantle. As a result, the forces of geomorphology are constantly altering the thickness of the earth's crust most everywhere, which is particularly noticeable in coastal areas, where the question of sea level has its relevance.

In short, because there is no system in place that can measure the direction or magnitude of the crust, scientists cannot evaluate how much of the ocean level change is due to a rise or fall of the ocean and how much is due to vertical movement of the crust. For Hampson, such arguments bring one to the realization that to have an absolute sea level at one location as compared to that of another location requires scientists to have credible methods for making such comparisons.

For millions of years, the earth has been subjected to successive cycles of active warming and cooling. These cycles were not of human origin, and they often reached temperatures much greater than those of the current period. Most of the science to explain these cycles is in its infancy, yet the climate science community has allowed its science research to be hijacked by the politics of money at the expense of the traditional scientific peer-review process, which is needed to understand earth's very complex climate systems and to create new paradigm shifts in human understanding of climate science.

16

The Solar Science behind
Climate Change

IN "WHAT Lies Beneath" (Patterson 2016), Matt Patterson begins
by taking the position that "new scientific discoveries in astrophysics
and archeology make the notion of 'settled science' risible." To make
his point, Patterson cites the wisdom of Donald Rumsfeld in stressing
the vital importance of unknown unknowns.

Once, an Indian mystic was explaining to an Englishman the structure of the universe. "The world sits atop a giant elephant," said the
holy man.

"That's all well and good," responded the Englishman with classic
Anglican sense, "but what does the elephant stand on?"

The wise-man's eyes widened, and he exclaimed, "Why, it stands
upon the shell of a grand and cosmic tortoise of course!"

"That's all well and good," again responded the Englishman, "but
what does the tortoise stand on?"

Surprisingly, this second question startled the fakir. Scratching his
head, he thought for a minute before replying with a single Hindi word
that may roughly be translated as "Something I know not what."

Patterson pictures the snide smirk on the Englishman's face having
cornered his interlocutor into admitting so much ignorance. Paterson
states, "After all, the English are heirs to a vast Western tradition the

sole ambition of which is to carefully categorize and explain the whole of nature."

Patterson proudly asserts that,

> this enterprise has been remarkably successful; certainly, Westerners can claim to have prodded and exposed a great many of nature's workings. In fact, we judge other cultures as modern to the extent and only to the extent that they adopt and appropriate Western scientific mores.

Yet Patterson argues that,

> the Western penetration of nature has been superficial at best. In fact, the foremost thinkers on the very edges of science are staring into an abyss of knowledge.

> As an example, Patterson cites that dark matter and dark energy combined (NASA 2018) make up 95 percent of everything, and yet we have no idea what these things are, how they work, or what they mean for the fate of the universe.

Patterson states,

> All of the stars and star-spawned material that is visible in the universe—including every planet, person or proton—accounts for a mere 5 percent of existence. It reflects radiation or emits it. It shines gloriously in the night sky, begging for our gaze and our awe. The rest? We know it's there, but we cannot see it.

In fact,

> dark energy is a mysterious force borne by something other than the photons that carry energy in our 5

percent world. We know it exists because we observe its effect on our luminous matter—the universe is flying apart. Gravity should be slowing the universe down and contracting its constituent parts; instead, things are flying apart at an increasing rate.

The long climb to scientific supremacy begun by Aristotle in his invention of symbolic logic has in the end taken us to the summit of what turns out to be a very small hill as we crane our necks upward at a looming, unseen, unending mountain range.

Patterson concludes,

> How profoundly unsettling it must be to realize perhaps we understand the smallest sliver of a story that was more vast and complicated than one could have ever imagined.

In the end, Western scientists may be forced when asked to explain what the vast majority of existence rests upon, to answer, "Something we know not what."

And despite the scientific mountains in the way, the long climb to scientific supremacy will continue advancing each step by paradigm shifts.

In their article "Ten Year Anniversary of the Climate Change Paradigm Shift" (Richman et al. 2013), Howard Richman and Raymond Richman cite Nobel Prize–winning economist Herb Simon, who explained the concept of a paradigm shift this way.

> The new paradigm begins with a new overall curve. Further research builds upon that curve by mapping the phenomena responsible for fluctuations from the curve. That's the normal scientific process. But establishing a new big data curve requires a paradigm shift.

According to Richmond and Richmond, who are focused on climate change, such a paradigm shift started in the 2000 decade when Israeli astrophysicist Nir Shaviv and Canadian geologist Jan Veizer published the groundbreaking study that laid out the chief long-term cause of climate change—cosmic rays. Richmond and Richmond provide support for this paradigm shift, which originated and is explained on Nir Shaviv's blog, "The Milky Way Galaxy's Spiral Arm and Ice Age Epochs and the Cosmic Ray Connection" (Shaviv 2006).

As noted by Richmond and Richmond, Shaviv had mapped the travels of the solar system through the spiral arms of our galaxy (shown in the top half of the draft shown in Shaviv 2006). Veizer had mapped the ice ages of the last 500 billion years (shown along with the fit to the cosmic ray inflow in the temperature record in the bottom half of this same graph). What they found is that ice ages occurred when the earth traveled through the spiral arms of our galaxy, periods when the earth must have been experiencing high levels of cosmic ray inflow.

According to Richmond and Richmond, other scientists had been laying the groundwork, but it was Shaviv's graph that caused the paradigm shift. At a time when the anthropogenic global warming paradigm was being promoted by scientists, his paper with Veizer was published in a geology journal, geology being the one scientific discipline that had never swallowed the man-centered view of climate change.

Geologists knew from the geological record that ice ages and greenhouse gases preceded humanity in the earth's history. Some also knew that carbon dioxide concentrations on today's earth are low compared to the levels during earlier epochs. They did not share the usual inflated view of humanity's power and importance.

When advocates of the anthropogenic paradigm challenged Shaviv's and Veizer's findings by claiming faulty data sets and faulty correlations, Shaviv responded in an article "Cosmic Rays and Climate" (Shaviv 2006) published in a physics journal in which he pointed out that his correlations work with two different signals of cosmic ray inflow and two different signals of earth's temperatures: "All four signals are consistent with each other, demonstrating the robustness of the link. If any data set is excluded, a link should still exist."

Richmond and Richmond noted that in recent years, more and more scientists have been exploring the new paradigm; they refer to the article "Cosmoclimatology: a new theory emerges" (Svensmark 2009) by Henrik Svensmark, one of its originators. As well, in 2009 Jasper Kirkby gave a lecture entitled "Cosmic rays and climate" (Kirkby 2009) at the CERN (Europe's premier research center) in which he reviewed the accumulating literature linking cosmic ray inflow variability with climate variability. One of Shaviv's and Veizer's graphs, cited in their article (Shaviv 2006), was featured prominently in that lecture.

In 2011, to provide empirical evidence to support the theory of cosmoclimatogy, Kirkby and his colleagues conducted an experiment at the CERN that provided a mechanism that could explain Shaviv's and Veizer's correlation. As outlined in their article "Role of sulphuric acid, ammonia and galactic cosmic rays in atmospheric aerosol nucleation" (Kirby et al. 2011), Kirby and sixty-two other authors found that ground-level cosmic-ray concentrations can increase the rate that cloud condensation nuclei form by a factor of between 2 and 10. Thus cosmic rays can cause clouds to form, which can reflect sunlight away from the earth, cooling the planet.

Richmond and Richmond argue that over the last decade, many studies have correlated cosmic ray inflow with earth temperatures whether the scale is hundreds of millions of years or hundreds of years. On shorter time scales, it turns out that high levels of solar activity (i.e., sunspots) are correlated with global warmth partly because an active sun blocks out cloud-initiating cosmic rays.

In "Hello Al Gore: Low Sun Spot Cycle Could Mean another 'Little Ice Age'" (Street 2015), Chris Street outlines the current known science to support the argument that the sun is the main driver of all weather and climate by first acknowledging that 99.86 percent of the mass in our solar system is the sun and accepting the recent scientific evidence that this great ball of violent fire in the sky has recently gone quiet in what is likely to be the weakest sunspot cycle in more than a century and actually flatlined (Dorian 2015) at the time he publishing his article. Street argues that known solar science shows that weak solar cycles such as the current one have been associated with benign "space weather" that can cause a "Little Ice Age" (Steele 2014).

Street notes that the strong solar activity impact on the earth has been scientifically studied since 1759 because it is known to have a direct impact on temperature changes in the "Thermosphere" (UCAR 2008a), which extends from about 56 miles to between 311 and 621 miles above the earth. The thermosphere sits between the Mesosphere (UCAR 2008b) below and Exosphere (UCAR 2008c) above. As the biggest layer of earth's atmosphere, Street notes that temperatures have been documented to rise and fall in the thermosphere depending on the amount of highly energetic solar radiation released from flares and sunspots.

According to Street, the Solar Cycle 24 has also experienced an unusually deep solar minimum that lasted from 2007 to 2009. That period had more spotless days on the sun than in any minimum for almost a century. In 2015, six years into Solar Cycle 24, the sun was virtually "blank" of sunspots (Dorian 2015).

For Street, weak solar cycles have been associated with benign space weather in recent times with geomagnetic storms that are generally weaker than normal. This may explain why since the extremely active hurricane seasons of 2004/05 there haven't been any category 3 to 5 major hurricanes to make US landfall at that strength. By all earth-based measures of geomagnetic and geoeffective solar activity, Cycle 24 has been extremely quiet.

But more important for Street, weak solar activity for a prolonged period of time can have a cooling impact (Dorian 2015) on global temperatures in the troposphere (UCAR 2011), which is the bottommost layer of earth's atmosphere, where we all live.

Street argues that there have been two notable historical periods with decades-long episodes of low solar activity. The first period is known as the Maunder Minimum (Wikipedia, MM) named after the solar astronomer Edward Maunder and lasting from around 1645 to 1715. The second one is referred to as the Dalton Minimum (Cregersen 2018), named for the English meteorologist John Dalton and lasting from about 1790 to 1830. Both historical periods coincided with colder-than-normal global temperatures and are often referred to by scientists as the "Little Ice Age" (Steele 2014).

For Street, the current historically weak solar cycle is a continuation of the twenty-year downward trend in sunspot cycle strength that began in Solar Cycle 22. This would explain the recent ten- to fifteen-year pause in the rise of global mean temperatures (Trenberth et al. 2014). The current surface warming hiatus is manifested by upper tropospheric teleconnection wave patterns that have created a climate anomaly (Liu et al. 2006), easterly trade winds, and large rainfall in the Pacific.

To provide added detail in support of the effect the sun has on the earth's climate, D. Bruce Merrifield ("Global Warming and Solar Radiation," Merrifield 2007), argues that without the impact of solar radiation, the temperature on the earth would be about the same as the temperature of space, which is about -454 degrees F. The amount of radiation reaching the earth is about 1,368 watts per square meter. This is a vast amount of energy that would require the simultaneous output of 1.7 billion of the largest power plants to match. About 70 percent of this solar energy is absorbed and 30 percent is reflected. However, Merrifield clarifies that the amount of solar energy reaching the earth is not constant; it varies in several independent cycles of different degrees of magnitude that may or may not reinforce each other.

Further, Merrifield explains that these cycles include a 100,000-year cycle that results from the elliptical orbit of the earth around the sun, a 41,000-year (obliquity) cycle that results from the tilt of the earth on its axis, a 23,000-year cycle that results from climatic precession or changes in direction of the earth's axis relative to the sun, and an 11-year sunspot cycle during which solar radiation increases and then declines.

According to Merrifield, each 100,000-year peak in radiation appears to last about 15,000 to 20,000 years, and each has been coincident with massive surges of carbon dioxide and methane (the greenhouse gases) into the atmosphere causing deglaciation of the polar and Greenland ice caps. Surges of these greenhouse gases have always been vastly greater than the amounts currently being generated by burning fossil fuels. As well, the most recent 100,000-year cycle raised sea levels four hundred feet in the first 10,000 years, but since then, sea levels have risen very little. In the current warming period, sea levels are rising only about three millimeters per year, and temperatures over the last 100 years have risen a modest 0.6 degree C.

Merrifield further explains that superimposed on this latest 100,000-year peak have been six secondary warming periods each coincident with additional surges of carbon dioxide and methane that lasted about 200 years and then subsided. Each of these previous warming periods was warmer than the current warming period, and current temperatures are below the median for the last 3,000 years.

Further, Merrifield states, it is remarkable that civilization first emerged in the Tigris, Euphrates, and Nile River Valleys about 3400 BC in a period of great warming. Even more remarkable, each of these secondary surges of greenhouse gases (none of human origin) has also been coincident with the rise of a major civilization.

As an example, Merrifield notes that three thousand years ago in the 1000 BC warming period, the Babylonian era emerged. Then five hundred years later, the Greek civilization flourished followed by the Romans four hundred years later. A thousand-year cold period followed through the Dark Ages, but then in the very warm Medieval Warm Period, the ice and snow melted on Greenland; the Danes farmed there for two hundred years until it froze over again.

17

The Science behind Life-Supporting CO$_2$ Greenhouse Gas

IN A speech delivered (Evans 2011) by Dr. David Evans in 2011, he opened with the following comment.

> The debate about global warming has reached ridiculous proportions. It is full of micro-thin half-truths, misunderstandings, and exaggerations. I am a scientist. I was on the carbon gravy train, I understand the evidence, I was once an alarmist, but now am a skeptic. Watching this issue unfold has been amusing but, lately, worrying. This issue is tearing society apart, making fools and liars out of our politicians.

> The idea that carbon dioxide is the main cause of the recent global warming is based on a guess that was proved false by empirical evidence during the 1990s. But the gravy train was too big, with too many jobs, industries, trading profits, political careers, and the possibility of world government and total control riding on the outcome. So rather than admit they were wrong, the governments, and their climate scientists,

> now outrageously maintain the fiction that carbon
> dioxide is a dangerous pollutant.

Evans fully accepts that carbon dioxide is a greenhouse gas and that the more carbon dioxide in the air, the warmer the planet. However, for Evans, the issue is not whether carbon dioxide warms the planet but by how much it does.

Most scientists on both sides of the argument also agree based on lab experiments for which the basic physics has been well known for a century on how much a given increase in atmospheric carbon dioxide raises the planet's temperature if just the extra carbon dioxide is considered. Evans argues that the disagreement comes with how the planet reacts to that extra carbon dioxide. Most critically, the extra warmth causes more water to evaporate from the oceans, which leads to the next issue: "Does this extra water hang around and increase the height of the moist air in the atmosphere, or does it simply create more clouds and rain?"

For alarmists, it is this increase moisture that is the core idea of every official climate model; they claim that for each bit of warming due to carbon dioxide, it ends up causing two extra bits of warming due to the extra moist air thus amplifying the carbon dioxide warming by a factor of three. So two-thirds of their projected warming is due to extra moist air (and related factors) while only one-third is due to extra carbon dioxide.

For Evans, there is simply no evidence for this amplification. To support his argument, he cites evidence provided by weather balloons that have been measuring the atmosphere since the 1960s, many thousands of them every year. During the warming of the late 1970s, 80s, and 90s, the weather balloons found no hotspot proving that the climate models are fundamentally flawed and that they greatly overestimate the temperature increases due to carbon dioxide.

There are now several independent pieces of evidence showing that the earth responds to the warming due to extra carbon dioxide by dampening that warming. Every long-lived natural system behaves this way and counteracts any disturbance; otherwise, the system would be unstable.

To provide a further understanding of the gases that make up the earth's atmosphere, their concentration, and whether they are permanent or vary in concentration, D. Bruce Merrifield ("Global Warming and Solar Radiation," Merrifield 2007) explains the characteristics of earth's three major elements identified as greenhouse gases: water vapor, carbon dioxide, and methane.

Water vapor is a greenhouse gas whose concentrations in computer models are currently not taken into account. Concentrations vary widely daily and over different sections of the earth. Low, thick clouds primarily reflect solar radiation and cool the surface of the earth. High, thin clouds primarily transmit incoming solar radiation but trap some of the outgoing infrared radiation emitted by the earth and radiate it back down to the earth warming its surface. The balance between cooling and warming is close, but cooling predominates.

Carbon dioxide entering the atmosphere has increased by about 1.8 percent per year since preindustrial times rising from about 280 ppmV to about 400 ppmV now—the highest in 160,000 years. However, preindustrial temperatures were much higher than are current temperatures, when carbon dioxide concentrations were at the much lower 280 ppmV.

Merrifield states that,

> Massive amounts of this gas are absorbed in the oceans, in terrestrial systems, and in the atmosphere with a relatively labile equilibrium between them. Concentrations in the oceans are about sixty times greater than in the land and atmosphere and about twenty times greater than in the atmosphere. Any warming of the oceans could release significant quantities of this gas.

Methane is more abundant in the atmosphere now than in the last 400,000 years, when concentrations were 278 ppmV (Ferretti et al.

2005). Since the Little Ice Age, concentrations have increased from 700 to 1767 ppmV, but since 1988, they also have ceased to rise for unknown reasons (Biello 2006).

According to Merrifield,

> Some 317 million cubic feet of methane are stored in US hydrates and some 49,000 quadrillion cubic feet exist in the world compared with known US natural gas reserves of ("only") 187 million cubic feet. World stores are 10 million teragrams of trapped methane vs. 5000 teragrams in the atmosphere. Enormous quantities of methane molecules are trapped in cage-like structures with water molecules on the ocean floor. Seismic shifts have been known to release large amounts of methane.

> For example, a deadly cloud of dissolved carbon dioxide and methane gas was released from Lake Nyos, Cameroon, in East Africa, that killed 1,700 people by a convective magmatic eruption that displaced the lower layer of the stratified lake in a volcanically active basin.

> However, methane may be primarily formed by bacterial degradation of vegetation.

Andre Lofthus, a professional scientist and physicist with over forty years of experience in aerospace and with extensive knowledge of atmospheric physics, wrote "Global Warming and Settled Science" (Lofthus 2014). In it, he argues, "Atmospheric transmission measurements taken in the 1950s demonstrate conclusively that increasing carbon dioxide concentration in the atmosphere cannot be the cause of the current man-made global warming."

For Lofthus, real scientists would demand to know the physics behind how increased carbon dioxide in the atmosphere causes global warming. Is there any real physics behind this unsupported bold assertion? Lofthus explains that based on test data from the 1950s, there is not.

To back up his bold assertion, Lofthus identifies three points based on physics and atmospheric physics. His first point is,

> molecules in the atmosphere absorb light waves over what are called spectral bands. The spectral band can be narrow, as small as a single wavelength, or broad, covering a continuum of wavelengths or frequencies. This molecular absorption causes increased vibration within the molecule exciting certain vibration modes. The physics of each molecule determine which wavelengths can be absorbed to excite internal vibrations. Spectral band absorption in the atmosphere can be quantified based on measurements over a certain distance through the atmosphere such as "90 per cent absorption in this spectral band over a distance of 300 meters at sea level through the atmosphere."

Lofthus's second point is,

> objects like the earth emit a spectrum or wavelength continuum of radiation that is completely described by Planck's Law (Planck 1900) of black-body radiation derived in the 1900s by Nobel-winning physicist Max Planck. That curve predicts the peak intensity of light from the sun in the visible spectral band and the peak intensity of light emitted by the earth in the Long Wavelength Infrared (LWIR) spectral band. Planck's curve has been validated by experimental data for over a hundred years, and it was a huge breakthrough for the physics community in the 20th Century.

Lofthus's third point is that,

> there are two spectral bands in which the carbon
> dioxide molecule absorbs infrared radiation. The first
> band is in what is called the Medium Wave Infrared
> (MWIR) spectrum, and the second spectral band is
> in the LWIR spectrum. Both bands are created by
> absorption of energy in a carbon dioxide molecule to
> excite stretching and/or bending modes of vibration
> within the molecule. The MWIR band of absorption
> excites stretching vibration modes, and the LWIR
> band of absorption excites bending vibration modes.

Lofthus provides a more technical explanation of the atmospheric transmission science (which can be reviewed in his above cited article Lofthus 2014) and refers his readers to a reference book published by the Office of Naval Research, a department of the US Navy titled *The Infrared Handbook* (Wolfe, et al. 1985), first published in 1978. According to Lofthus, this book contains atmospheric transmission data at sea level based on measurements that were taken in the 1950 time frame, much before any recent increases in the concentration of carbon dioxide in the atmosphere.

Lofthus concludes that based on this established science, increasing the concentration of carbon dioxide in the atmospheric mixture of gases by burning fossil fuels or by bovine flatulence will not increase the measured absorption in the carbon dioxide LWIR band above the 100 percent level that was measured and reported in *The Infrared Handbook*. You cannot get more than 100 percent absorption. And yet that appears to be the basis of the theory of "man-made" global warming.

To provide added evidence to support the effect the sun has on the earth's climate, D. Bruce Merrifield ("Global Warming and Solar Radiation," Merrifield 2007), argues that since the emergence of civilization and until relatively recently, economic prosperity has been primarily based on agriculture, and each of these warming periods has been accompanied by an improved climate for growing food. After each surge, subsequent cooling of the climate occurred with declines in greenhouse gasses and shorter growing seasons, perhaps contributing also to observed societal declines.

According to Merrifield,

> mutually supportive data documenting these episodes have accumulated from many sources. They include cores from the Antarctic ice cap, from the Sargasso Sea, from stalagmites, from ocean up-welling and from the shells of crustaceans trapped in pre-historic rock formations.

> For example, the geological record from some 55 million years ago documents a great warming period, which occurred over a "geological instant." Carbon dioxide surged to about 1000 ppmV, and temperatures rose 5 to 7 F. higher than current global temperatures [Science 2005]. Methane also increased dramatically in this and other warming periods, with de-glaciation following each warming period.

> Recently, an analysis of ice cores from the Antarctic ice cap [Science 2006] have shown that over the last half million years, there have been sudden and repetitive powerful surges of carbon dioxide into the atmosphere about every 100,000 years, with rapid de-glaciation, followed then by re-glaciation in long cooling periods. We now are at the latest of these peaks, which terminated the last ice age and raised sea levels about 400 feet.

Merrifield notes that this Antarctic data indicate that the rise in carbon dioxide, then leveled off for unexplained reasons. Also, net increases of both carbon dioxide and methane have now ceased since 1988, although the production of human-generated carbon dioxide, methane and nitrous oxide continue to accelerate.

So if carbon dioxide is not the single most important forcing agent as claimed by the alarmists, what is causing the natural warming and cooling events that the earth has experienced over the last 500-plus million years? What are the other forcing agents affecting the earth's climate and its daily weather patterns?

Taking up this challenge of identifying other forcing agents affecting the earth's climate, John Kudla ("Things Your Professor Didn't Tell You About Climate Change," Kudla 2018) starts by argues that the sun as discussed in the previous chapter has a larger effect on the earth's climate than any other forcing agent. Small changes in solar insolation due to variations in the sun's energy output or cyclical variations in the earth's orbit, known as the Milankovitch Cycles (Indiana 2018), can make a big difference in the earth's surface temperature. Kudla clarifies that other agents affect earth's climate and its surface temperatures including greenhouse gases, ocean currents, volcanic eruptions, and many other things, but the sun is still the eight-hundred-pound gorilla in the room.

Kudla states that over the last 450,000 years, the normal average global temperature has been approximately 5 degrees C cooler (Interglacial Comparisons 2014, ClimateConcern 2014) than it is today. During that time, earth's climate cycled (Climate Change and Human Evolution 2013, Palomar 2013) between long cool periods known as glacials, which could last 50,000–100,000 years, and shorter warm periods called interglacials, which usually lasted 10,000–20,000 years. During glacial periods, glaciers and continental ice sheets develop and grow. During interglacial periods such as the one being experiencing now, the earth warms and the sea level rises as most of the ice melts.

Kudla explains that due to the above-mentioned factors and other natural climate oscillations such as the El Niño Southern Oscillation (ENSO), the eleven-year sunspot cycle, the Pacific Decadal Oscillation

(PDO, Spencer 2008), the Atlantic Multidecadal Oscillation (Wikipedia, AMO), and perhaps the De Vries solar cycle, the world's climate continually changes.

In an online posting "The Pacific Decadal Oscillation (PDO): Key to the Global Warming Debate?" (Spencer 2008b), Roy Spencer defines the Pacific Decadal Oscillation (PDO) as an internal switch between two slightly different circulation patterns that occurs every thirty years or so in the North Pacific. It was originally described in 1997 in the context of salmon production. It has a positive (warming) phase that tends to warm the land masses of the northern hemisphere as well as a negative (cooling) phase.

Like the El Niño and La Niña oscillations of the tropical Pacific (also called the El Niño Southern Oscillation, ENSO), the PDO represents two different average circulation states that the ocean-atmosphere system seems to have a difficult time choosing between. But whereas the ENSO changes every few years, the PDO changes every thirty or so years. This long-time scale makes the PDO a potential key player in climate change.

One of those oscillations occurred between 1940 and 1977 as the earth went through a minor cooling trend possibly linked to the PDO. This prompted a global cooling scare as scientists feared we were sliding into another glacial period ("1970s Global Cooling Alarmism," Santayana 2013). Kudla argues that this means the present warming trend could be a natural climate oscillation unrelated to carbon dioxide or possibly a combination of both.

In "The Looming Threat of Global Cooling" (Easterbrook 2010), geologist Don Easterbrook also attributes natural climate variations to solar irradiance and deep ocean currents. He notes the undeniable link between the PDO shifting to its warm mode in 1915 and 1977 with global warming resulting both times. Conversely, in 1945 and 1999, the PDO moved to its cool mode and the globe cooled right along despite a rapid increase in atmospheric carbon dioxide during the period. Climate changes in the geologic record show a regular pattern of alternate warming and cooling within a 25- to 30-year period for the past 500 years.

Given that these observations prove out, Easterbrook concludes that we should "expect global cooling for the next 2–3 decades that will be far more damaging than global warming would have been."

Easterbrook further notes a strong correlation between PDO and solar activity as did geophysicist Victor Manuel Velasco Herrera ("Scientist Warn that Earth will Enter 'Little Ice Age' Due to Decrease in Solar Activity," Herrera 2008), who believes an even longer cold spell (sixty to eighty years) has begun triggered by a decrease in solar activity. Habibullo Abdussamatov ("Top Russian Scientist Predicts 100 Years of Global Cooling," Abdussamatov 2011) agrees; he illustrates how the eighteen Little Ice Ages that occurred in the past 7,500 years can all be attributed to "natural bicentennial variations in the average annual values of the total solar irradiance (TSI)" and its secondary subsequent feedback effects (natural changes in the albedo (reflection coefficient), water vapor abundance, etc.).

Abdussamatov demonstrated that each time the TSI reached a peak (up to 0.2 percent), a period of global warming began "with a time lag of 15 ± 6 years defined by the thermal inertia of the Ocean (despite the absence of anthropogenic influence)." Conversely, "Each deep bicentennial descent in the TSI caused a Little Ice Age." Based on the present cycle, the astrophysicist had expected "the beginning of the new Little Ice Age epoch approximately in 2014."

Since the alarmists are now claiming extreme weather is the result of climate change, the late Dr. William M. Gray, a renowned hurricane specialist, published "Global View of the Origin of Tropical Disturbances and Storms" (Gray 1968). His research focused on most of the warming experienced in the past thousand years and found that it could be attributed to deep ocean circulations strengthened and weakened by century-scale salinity variations. While the relationship of sea surface temperatures to evaporation, rainfall, and wind patterns, albedo, and ultimately air temperature is complex and beyond the scope of his research, which translates to ocean-, not carbon dioxide-driven global temperatures.

Gray believes the Medieval Warm Period (MWP) was a result of a multicentury slowdown of the North Atlantic Meridional Overturning Circulation (MOC), a global circulation cell wherein surface waters in the high latitudes are cooled thereby becoming denser; this dense water sinks and flows toward the equatorial regions. Gray argues that was similar to that experienced in the twentieth century and corresponded to similar warming. Conversely, the Little Ice Age was a period of stronger than average MOC as we are beginning to see today. Gray also predicts that strengthening ocean currents portend global cooling over the next few decades even as carbon dioxide levels continue to climb.

Despite IPCC modelers declaring the probable climate sensitivity or the amount of warming to be expected by a doubling of atmospheric carbon dioxide this century to be 3 degrees C and attribute two-thirds of that figure to positive feedback from clouds, Gray argues that clouds actually provide .5 degree C of negative feedback for a total climate sensitivity of .5 degree C.

Indeed, the designation of clouds as negative rather than positive feedback has been a lesson taught by Spencer and MIT's Lindzen for years. Spencer argues that the IPCC modelers derive their catastrophic warming predictions from algorithms whereby carbon dioxide warming causes a decrease in clouds, which lets in more sunlight and leads to more warming. Spencer argues quite the opposite—weak warming increases clouds letting in less sunlight, which leads to less warming. In fact, Spencer believes that a full 75 to 80 percent of warming could be due to cloudiness changes due to the PDO. In other words, most of past warming is likely natural and climate sensitivity is likely closer to 0.5 degree C.

In "Professor Richmond Lindzen's Congressional Testimony" (Watts 2010a), Lindzen fundamentally agrees with the 0.5 degree C figure noting that a doubling of carbon dioxide by itself contributes only about 1 degree C to greenhouse warming.

> We see that all the models are characterized by positive feedback factors (associated with amplifying the effect of changes in CO_2), while the satellite data implies that

> the feedback should be negative. Only with positive
> feedbacks from water vapor and clouds does one get
> the large warmings that are associated with alarm.
> What the satellite data seems to show is that these
> positive feedbacks are model artifacts.

Frequent *Watts Up With That?* contributor Willis Eschenbach outlined his hypothesis in "Willis publishes his thermostat hypothesis paper" (Watts 2010b) in which he explains that temperatures are kept within a narrow and fixed range by a governing mechanism of clouds and thunderstorms set by the physics of the wind, waves, and ocean but not carbon dioxide forcing. Past IPCC expert reviewer Tom V. Segalstad also insists that clouds are the real thermostat with far more temperature-regulating power than carbon dioxide ("Web info about CO_2 and the asserted 'Greenhouse Effect' Doom," Segalstad 2018).

In *Heaven and Earth: Global Warming: The Missing Science* (Plimer 2009), geologist Ian Plimer notes another problem with IPCC models—they completely ignore the role of volcanoes in their analysis. Neither terrestrial volcanoes, which expel heat, water vapor, and carbon dioxide, nor submarine volcanoes, which add heat and gases to the oceans and increase its levels of carbon dioxide play a part whatsoever in IPCC predictions.

Despite all the doomsday predictions of AGW being issued by climate alarmists as a result of increased emissions of carbon dioxide from the combustion of fossil fuels in the twenty-first century, S. Fred Singer ("Cause of Pause in Global Warming," Singer 2014) notes that there has been essentially no global warming since 1998. Some would choose 1997 while others would more conservatively use 2002 as the proper starting date based on satellite data. The existence of the global warming hiatus is creating a scientific challenge for climate skeptics and a real crisis for alarmists and cannot be ignored by those who consider themselves to be scientists or by responsible politicians.

Singer then offers a number of possible causes to explain this global warming hiatus or pause. First, Singer suggests that the pause is simply a statistical fluctuation like tossing a coin and coming up with fifteen heads in a row. Such an explanation cannot be dismissed out of hand even though the probability of that happening becomes less with each passing year of no global warming.

Singer then distinguishes between internal and external causes that might offset the expected global warming from carbon dioxide. Internal causes rely on negative feedback from either water vapor or clouds, which act to decrease the warming that should be attributed to increasing carbon dioxide. The problem with internal effects is they can never fully eliminate the primary cause almost by definition. So even if they diminish the carbon dioxide effect somewhat, there should still be a remaining warming trend though small.

In the case of water vapor, to obtain empirical evidence, scientists would look to see if the cold upper troposphere was dry or moist. If moist as assumed implicitly in the AR5 IPCC General Circulation Models (GCM), there is positive feedback with an amplification of warming caused by CO_2. On the other hand, if the upper troposphere is dry, most emissions into space take place from water vapor in the warm boundary layer in the lower troposphere. This leaves less energy available to be emitted into space from the surface through the atmospheric window and therefore produces a cooler surface.

The research Singer cites to support the negative water vapor feedback is based on the work of William Gray (Gray 1968, cited above), who pictured cumulus clouds carrying moisture into the upper troposphere but occupying only a small area; the remaining (and much larger) area experiences descending air (subsidence)—hence drying. In principle, it should be possible to measure this difficult-to-explain effect easily by using available satellite data.

Singer further argues that negative feedback from increased cloudiness is easier to describe but more difficult to measure. The idea is simply that a slight increase in sea surface temperature (from the greenhouse effect of a rising carbon dioxide) also increases evaporation (according to the well-known Clausius-Clapeyron relation, MIT

2018) and that this increased atmospheric moisture can also increase cloudiness. The net effect is a greater albedo (reflected sunlight) and less sunlight reaching the surface and therefore negative feedback that reduces the original warming from increasing carbon dioxide.

Unfortunately, establishing the reality of this cloud feedback requires a measurement of global cloudiness with an accuracy of a small fraction of a percent, a very difficult challenge.

Singer then turns his focus to the external effects that might explain the existence of a global warming pause, the principal ones being volcanic and solar activity. The problem here is one of balancing; the amount of cooling by volcanism for example has to be just right to offset the warming from carbon dioxide during the entire duration of the pause. It is difficult to picture why exactly this might be happening; the probabilities seem rather small. Still, the burden is on the proponents to demonstrate various kinds of evidence in support of such an explanation.

Similarly, atmospheric aerosols, generally human-caused, can increase albedo and cool the planet especially if they also increase cloudiness by providing condensation nuclei for water vapor.

Singer notes that all the explanations cited here act to reduce climate forcing, the energy imbalance measured at the top of the atmosphere.

Singer suggests that there is an important school of thought that does not rely on offsetting the forcing from increased carbon dioxide but instead assumes that there really exists an imbalance at the top of the atmosphere and that global warming is taking place somewhere but is not easily seen. Many scientists assume that the missing heat is hiding in the deep ocean. It is difficult to see how such a mechanism could function without also raising surface temperatures, but an oscillation in ocean currents might produce such a result as discussed above.

Singer challenges this theory by arguing that if measurements could demonstrate a gradual increase in stored ocean heat, one would be forced to consider possible mechanisms. Its proponents might be asked, Why did the storage increase start just when it did? When will it end? How will the energy eventually be released? And with what manifestations?

Singer cites yet another possibility worth considering: the missing energy might be used to melt ice rather than warm the ocean. Again, quantitative empirical evidence might support such a scenario. But how to explain the starting date of the pause? And how soon might it end?

With all the unsupported scientific hype generated by the climate alarmists predicting the end of life on earth because of increasing global temperatures to dangerous levels because of increasing carbon dioxide levels, there is almost no discussion as to the good that is happening from the increased concentration of carbon dioxide, a greenhouse gas in the earth's atmosphere.

In "Blessing or Curse? The Curious Case of Carbon Dioxide" (Jayaraj 2017), Vijay Jayaraj reminds his readers that carbon dioxide is an odorless, invisible trace gas in the atmosphere that acts as an important source of life for everything that lives on earth. In fact, plant and animal life on earth would be impossible without carbon dioxide.

Jayaraj explains that carbon dioxide is an integral part of the photosynthesis process. Plants synthesize carbon dioxide and water to produce chemical energy that sustains them. Together with water and sunlight, carbon dioxide acts as the elixir of life for the plant and animal kingdom.

Jayaraj notes that historical data suggests this increased carbon dioxide concentration has helped plant growth globally enabling civilizations to increase their agricultural output drastically. Record food crop outputs in recent decades are due in significant part to increased atmospheric carbon dioxide concentrations. Jayaraj cites as an example the increase in production of major crops such as wheat, rice, maize, and soybeans.

To make his point, Jayaraj cites "Greening of the Earth and it drivers" (Zhu et al. 2016), in which the authors captured relative change of leaf area index from 1982 to 2009 due to carbon dioxide fertilization and found that most parts of the world recorded an increase in plant canopy by as much as 14 percent primarily due to increase in atmospheric carbon dioxide concentration.

Further, Jayaraj cites a research paper "Quantification of the response of global terrestrial net primary production to multifactor global change" (Li et al. 2017) by Peng Li and colleagues. They found that net primary production—the productivity of individual or groups of plants—increased by 21 percent globally between 1962 and 2010. The increase in atmospheric carbon dioxide concentrations was found to be the dominant factor for this extraordinary increase in plant growth.

Jayaraj argues that there has been a 40 percent increase in atmospheric carbon dioxide concentration since the beginning of the Industrial Revolution. He further argues that world food production must increase by more than 70 percent from today's levels to meet global food demand in the year 2050 because of the growing global population. Without the benefits of carbon dioxide, the world food supply would fall short unless huge new tracts of land were converted from wilderness to farmland.

If anything, the carbon dioxide concentration in the atmosphere has helped life on earth and provided hope for the billions who live in poverty. It is time for the climate alarmists and the global liberal media to stop calling carbon dioxide a curse and start calling it a blessing. It is time to acknowledge the immense benefits that humans have enjoyed due to carbon dioxide.

Regardless of any unsettled science details, it seems sure that current climate models cannot represent what is actually happening in the atmosphere, and therefore, we should not rely on predictions from such unreliable models based simply on increases of carbon dioxide. It should be obvious that this discussion has important policy consequences since so many politicians are wedded to the idea that carbon dioxide needs to be controlled to avoid dangerous changes of the global climate.

18

Free-Market Economic Reality and Climate Change

SOCIALISM KILLS. From the former Soviet Union and Cuba to Venezuela, anywhere socialism has been tried, it has robbed people of freedom and property and has produced economic stagnation and misallocation of resources resulting directly or indirectly in millions of deaths.

In "Energy Socialism Kills" (Burnett et al. 2018), H. Sterling Burnett and Justin Haskins argue that energy socialism as touted by a young, self-described socialist Alexandria Ocasio-Cortez elected as a Democrat to the House of Representatives in New York in the 2018 midterm elections would be just as deadly. Her Green New Deal states, "Climate change is the single biggest national security threat for the United States and the single biggest threat to worldwide industrialized civilization." Ocasio-Cortez proposes a plan to transition the United States to a 100 percent renewable energy system by 2035 and to totally restrict its citizens' freedoms, prosperity-enhancing infrastructure, and institutions.

Burnett and Haskins argue this Green New Deal would result in a massive transformation of the US energy system, which took more than eighty years to build. For Burnett and Haskins, it is the extent to which socialist thinking has captured the Democratic Party's imagination with hundreds of Democrats signing a pledge to push for this 100 percent renewable energy makeover.

Even worse, dozens of Democratic lawmakers are attempting to pass radical climate legislation that they're calling "the most aggressive piece of climate legislation ever introduced in Congress." The Off Act (Schlosberg 2017) would require "100 percent renewable energy by 2035 (and 80 percent by 2027), places a moratorium on new fossil fuel projects, bans the export of oil and gas, and also moves US automobile and rail systems to 100 percent renewable energy." These policies supported fully by numerous environmental extremists would destroy millions of jobs and put the US at a huge disadvantage when competing against other countries especially China, India, and other nations whose environmental laws are much less stringent than those of the US.

For Burnett and Haskins, these policies would impose a form of environmental socialism. Instead of finding ways to make wind, solar, and other forms of renewable energy technologies cheaper and more competitive with conventional energy sources, socialists such as Ocasio-Cortez and liberal left members of Congress want to use the government to destroy whole industries composed of hardworking Americans and then take trillions of dollars from taxpayers to keep failing renewable energy companies afloat. They could not care less that they would put honest people out of work, drive up energy costs, and hurt the poorest citizens more than any other group.

Burnett and Haskins argue that the US economy is the envy of the world; it is built on a power system reliant on relatively inexpensive and reliable energy sources primarily and necessarily fossil fuels. However, energy socialism gained a larger foothold in the electric power market when federal and state governments began providing lavish subsidies, tax credits, and tax abatements to politically connected, big green solar and wind energy companies.

Many states compounded this grave error by mandating that utilities operating in their borders ensure that ever-growing percentages of the electricity they provide come from select renewable energy sources.

People in states with renewable power mandates have since seen their electricity bills rise by a greater amount compared to those living in states without renewable power diktats.

Burnett and Haskins cite the state of Europe, which is much further down the road to energy socialism than most of the US, where thousands of people die in winter due to a lack of reliable heat, and where during the summer, more die from not having access to reliable air conditioning. European politicians react to these problems by telling their fellow citizens that they will have to make do with less and plan for shortages. Europe's energy problems did not result from some inability to produce energy; rather, they come from a decision by politicians to shutter reliable fossil fuel and nuclear power plants as part of their misguided push to fight climate change.

These are the kinds of Third World problems that come with energy socialism—less reliable power, higher energy costs, greater poverty, massive job losses, and lower economic productivity. Socialism can't fix our problems, but it sure can make things a lot worse.

In "Destroy Capitalism, Save the Climate" (Ellison 2018), Robert Ellison cites the film *This Changes Everything—The Film* (Lewis et al. 2014), which outlines the policies from global warming alarmists who present tales of the collapse of Western civilization and capitalism leading to less growth, less material consumption, less carbon dioxide emissions, less habitat destruction, and a late chance to stay within the safe limits of global ecosystems.

Ellison acknowledges the progressives are right about their tale that economies are fragile when market movements can be fierce and recovery glacially slow sometimes. However, Ellison argues that these economic problems are not intrinsic to free markets and their interaction with the environment. The great majority of these problems are created by poor judgment or blundered into through stupidity. The scenario most concerning for Ellison is where markets are deliberately destabilized to hasten the end of capitalism. The following examples are suspiciously the objective of the climate warming alarmists: creeping

tax takes, overspending by government, printing money, keeping interest rates too low for too long or too high for too long, penalty taxes on
primary inputs, and implementing market-distorting subsidies. All of
these examples fall into the category of stupid policies.

For Ellison, the rational management of economies requires interest rates to be managed through the overnight cash market to restrain
inflation to the 2 to 3 percent range. Markets need fair, transparent,
and accessible laws including open and fair markets. The optimal tax
take is some 23 percent of GDP, and budgets are balanced. Markets
operate best in a robust democracy. These nuts and bolts of market
management—mainstream market theory and practice pioneered by
F. A. Hayek (Wikipedia, Hayek)—keep economies on a modest and
stable growth trajectory as much as possible.

Further, Ellison argues that this economic growth provides resources for solving problems such as restoring organic carbon in agricultural soils, conserving and restoring ecosystems, better sanitation
and safer water, better health and education, updating the diesel fleet,
and other productive assets to reduce the strong climate effects of black
carbon, and developing better and cheaper ways of producing electricity. Over time, free-market economies create the resources to develop
low-cost alternatives to the fossil fuels that are increasing in scarcity
and cost.

Ellison agrees fully that there is only one planet and that it is important to protect the environment. Our concern for it extends well
beyond carbon dioxide. Population, development, technical innovation, multiple gases and aerosols across sectors, land-use change, and
the environment are the broader context.

For Ellison, the solutions to the multiple problems of people in
the world are both simpler and more complex than overthrowing democracy and free markets. The energy solutions are technological—
primarily gas to a more advanced nuclear strategy ("Greenhouse gas
solutions," Terra and Aqua 2015)—if nothing better comes along.
Management of the global commons is a messy and complex human

problem requiring the most modern theories and models of human behavior that are well beyond the simplistic response of caps and taxes on energy. Environmental management involves the strategic deployment of methods and technologies across landscapes, industries, and infrastructure, and implementation requires a different approach from top-down government regulation, which is failing business and the environment.

To provide a real-world context to Ellison's argument that top-down government regulation is the wrong approach to environment management, Mark Ahlseen provides an overview ("The Cost of the United Kingdom's Energy Policy," Ahlseen 2018) of the first two decades of this century in which the UK adopted an energy policy designed to reduce the use of fossil fuels and especially coal to generate electricity and replace them with renewables especially wind.

As a resource, Ahlseen used a review of the UK's energy policies ("Cost of Energy Review," Helm 2017) completed in 2017 by Dieter Helm, professor of energy policy at the University of Oxford. It is Ahlseen's opinion that citizens and governments around the world should take some lessons from Helm's work.

Ahlseen argues that most economists acknowledge the benefit of unfettered supply and demand in determining the allocation of scarce resources. However, many people think government must intervene with regard to some commodities. Electricity is one.

Ahlseen notes that there are three stages in getting electricity to consumers: generation, transmission, and distribution. Unlike most products, electricity must be generated in a way that meets the instantaneous demands of consumers. Since current technology does not allow for storage of electricity during low demand, generating capacity must at every moment be sufficient to meet peak demand.

Ahlseen acknowledges that a for-profit company would not invest in excess capacity; instead, its generating plants would meet average, not peak demand. For this reason, there is the argument for government regulation. If however an electricity firm could store excess electricity generated during low demand and transmit it during high demand, most economic regulation would become unnecessary.

However, environmental regulation is a different matter. Ahlseen notes that in the UK, as in most developed countries, environmental regulations have already succeeded in reducing traditional pollutants from electricity-generating plants—soot, sulfur oxides, nitrous oxides, carbon monoxide, and heavy metals—to levels so low as to be of little or no risk to health. Practically, the only things coming out of the smoke-stacks of most coal and natural gas electric generating plants are water vapor and carbon dioxide.

However, climate alarmists now consider carbon dioxide a pollutant, the emissions of which are thought to drive dangerous global warming, which makes it the new focus of environmental concern with electricity generation.

Ahlseen argues that to address this carbon dioxide pollutant issue, the UK passed the Climate Change Act (UK Gov 2008) in 2008 aiming to achieve an 80 percent reduction compared with 1990 levels in carbon dioxide emissions by 2050. Multiple policies have been introduced to achieve this goal, but many critics, Helm among them, argue that some of these regulations are contradictory, redundant, or superfluous.

Helm does not oppose carbon dioxide emission reduction, but he believes that it should be achieved in the least costly fashion. However, Helm estimates that regardless of what the benefits of emission reduction are, the government policies have wasted around £100 billion ($132 billion) to date, and the waste will continue to rise unless reforms are made. Helm is particularly concerned that the government picks which new technologies to back leading to regulatory capture by special interests. Regulatory capture denotes a situation in which a government agency meant to serve public interests serves the interests of affected commercial or political entities instead. Helm notes,

> Government has got into the business of "picking winners." Unfortunately, losers are good at picking

governments, and inevitably—as in most such picking-winners strategies—the results end up being vulnerable to lobbying, to the general detriment of households and industrial customers.

Helm rightly notes in his review that technological changes are occurring at a rapid pace, but for Ahlseen, the environmental bureaucracy is not at all different from any other governmental bureaucracy with the billions being wasted only growing as new technologies slow to be adapted when they impinge on the vested interests' current benefits.

Ahlseen projects that the decarbonizing of the UKs energy production will come at a cost. A rough estimate of its cost to the UK as a whole can be calculated to be roughly £15.8 billion in the year 2020 alone and £292.68 billion cumulatively through 2030. Ahlseen speculates as to what other social priorities could be met with this money.

In another example of potential government interfering in the natural workings of the free-market system due to climate alarmism, H. Sterling Burnett ("The Carbon Tax Rebate Scam," Burnett 2017a), cites a group of old-guard swamp Republicans calling themselves the Climate Leadership Council (CLC) that had joined climate alarmists including failed Democratic Party presidential candidate Gore in calling for a tax on carbon dioxide emissions. The group claims increasing carbon dioxide emissions poses a threat for earth's people, animals, and plants even though Congress rejected a carbon tax resolution in 2016.

Burnett argues that economic analyses of various carbon tax proposals consistently show they would harm all Americans, would be detrimental to the US economy, and would burden businesses with unnecessary costs making them less competitive in the global marketplace. For Burnett, a stand-alone tax directed at carbon dioxide emissions would have a disparate impact on the poorest Americans because they spend a greater portion of their incomes on energy and energy-intensive products compared to upper-middle-class and relatively wealthy Americans. The CLC acknowledged this, so it promised to rebate the revenue generated by the tax back to the public.

The CLC's plan would begin with a carbon tax rate of $40 per ton. At that rate, it estimates that a family of four would receive approximately $2,000 as part of its carbon refund in the program's first year, and it brags that the bottom 70 percent of Americans under its plan would receive more in refunds than they would pay in increased energy costs, but they completely ignore the negative effect such a tax would have on the 30 percent of Americans who would pay more than their fair share. In other words, it's another soak the rich and subsidize the poor entitlement scheme.

According to Burnett, what is not clear in the implementation of carbon taxes is their cost to administer. Burnett asks how the government can calculate or track the amount people would pay in new energy taxes to qualify for their rebates. Would everyone have to receive a national carbon tax ID card, or would people have to use their Social Security cards when they purchased gasoline or paid their electric bills? Whichever method chosen, the government needs a way to determine who gets which rebate, and that will likely (and rightfully so) raise the hackles of people who oppose the creation of a national ID.

Burnett speculates that Americans might be required to itemize the carbon taxes they pay, meaning taxpayers would have to keep track of even more tiny receipts to properly file their tax returns. Just as with every other government program, there would be huge costs associated with collecting, tracking, auditing, and archiving the taxes paid and rebates paid out.

The only way to understand the unquestioning acceptance of the climate alarmism infecting free-market economies around the world, governments implementing expensive and unreliable energy sources, and the financial handicapping of their citizens can best be explained by the expression "Follow the money" made popular in the 1970s Watergate movie *All the President's Men* as a means of tracing a path of corruption.

In "Follow the Money on Climate Caterwauling" (Joondeph 2018d), Brian C. Joondeph argues that in the case of climate warming alarmism, following the money is a means of finding motivation behind government funding, the compromised climate science activists, and the complicit liberal big media.

Joondeph argues that when following the money from government funding and subsidies, one finds it heavily favors research promoting the theme that man-made global warming is real, preventable, and apocalyptic. This is "federal funding induced bias" as described by the Cato Institute in the article by David E. Wojick and Patrick J. Michaels, "Is the Government Buying Science or Support? A Framework Analysis of Federal Funding-induced Biases" (Wojick et al. 2015).

Joondeph explains that this bias occurs when a scientist publishes research findings contrary to the liberal left climate dogma and his or her funding dries up. A young climate scientist eager to climb the academic ladder, earn tenure, and have a successful career quickly learns how to promote an agenda rather than perform objective scientific research.

By contract, Joondeph argues the free markets allow businesses to follow their money without government interference based on the laws of economic guiding decisions and behavior. That is to say, the least restrictive markets with completely uninhibited free markets tend to coincide with countries that value private property, capitalism, and individual rights. Political systems that avoid regulations or subsidies for individual behavior necessarily interfere less with voluntary economic transactions. Additionally, free markets are more likely to grow and thrive in a system in which capitalists have an incentive to pursue profits. For Joondeph, free markets represent "Voluntary exchanges … characterized by a spontaneous and decentralized order of arrangements through which individuals make economic decisions."

Joondeph translates this by saying free markets are driven by return on investment or profit, and as a result, individuals and businesses will analyze a myriad factors and make economic decisions based on their analyses. Few businesses deliberately make decisions that will lose money with the exception of certain unnamed cable news networks.

It is obvious that climate science is one of the factors influencing business decisions. For example, opening a ski resort in Texas, a state that rarely sees snow or freezing temperatures, is not a wise business decision. On the other hand, since climate changes are small as in a snowier winter one year and a dry winter the next year, opening a ski resort in Colorado is not a climate-dependent decision as there will be snow every winter despite its being more or less in any given season. Longer-term climate changes are beyond the realm of most businesses, which are not concerned about the next ice age in 10,000 years.

One main reason business and the public are rejecting the climate fear-mongering strategy being propagated by climate alarmists and the liberal big media is that none of their end-of-world predictions have materialized.

For Joondeph, free markets live in the real world and do not wear ideological blinders and don't have destructive self-interests. To make his point, Joondeph notes that Miami is one of the cities the UN claims "will be drowned by global warming." Yet in Miami (Wallman 2018), "A gargantuan new shopping mall and theme park—the largest in North America—will be built in South Florida." This $4 billion project will comprise 2,000 hotel rooms, 2,000 apartments, 2 million square feet of retail and office space, and an indoor ski hill. Joondeph asks, "What property developers and investors would risk $4 billion on a project soon to be underwater?" Miami is only six feet above sea level, far less than the sea level rise Gore predicted.

In another example, in "South Florida Real Estate Boom Not Dampened By Sea Level Rise" by Greg Allen (Allen 2017), NPR doesn't follow the money either. The article signs onto climate alarmism.

> For coastal communities from Florida to Texas, this year's hurricane season may be a preview of what's to come. Scientists say with climate change, in the future we're likely to see more severe hurricanes and heavier rain events. In addition, as ice sheets melt, sea levels

are rising faster, flooding low-lying coastal areas such as Miami.

Despite these misguided predictions, free-market economics tells a different story. Quoting from the same NPR article,

> But in South Florida, the dire predictions have done little to dampen enthusiasm for development. In Miami's urban core, there are some 20,000 condominium units in various stages of completion.

Joondeph sarcastically states that this action by the free market seems pretty foolish in defying the predictions of climate alarmists—building in a city soon to be underwater—unless the builders know or believe otherwise by not buying into climate doomsday prophesies.

Joondeph cites an even more interesting example of free-market skepticism involving the liberal leftist Hollywood elites who have signed on to the global warming hysteria. This website boasts a list of "Celebrities Who Live(d) in Malibu" (Ranker 2018) and on the beaches of Malibu soon to be underwater if the seas rise by twenty feet—Leo DiCaprio, Sean Penn, Ellen DeGeneres, Tom Hanks, and others. Why aren't they worried about the climate alarmists' dire predictions? Add to this list of hypocrites is Gore, who in 2010 purchased his own ocean-front property ("Al Gore Buys $8.9 Million Ocean-view Villa," Beale 2010) not on the beach but in Montecito, California, quite close to the ocean. Joondeph wonders why he would not be worried about rising seas.

Joondeph argues that the public can listen to politicians most of whom have never held a real-world, nonpolitical job in their lives or who follow the money, not listening but watching actual businessmen and businesswomen, researching, analyzing, and concluding that climate hysteria is just that.

Do not get sucked into the feel-good, virtue-signaling causes designed for personal enrichment whether financial or political without regard to the scientific and business reality around you. Trust those who put their own money at risk. They sing a different song about global warming compared to the chicken littles. When in doubt, follow the money.

In another economic example of free-market pragmatism, no matter how hard environmental do-gooders are trying to kill coal, they're clearly not succeeding. The expression "Follow the money" is particularly appropriate when listening to the fear-mongering from climate alarmists in their efforts to kill coal as a global energy source. In a *Wall Street Journal* article entitled "Coal Shows Resilience in Global Comeback" (Salvaterra 2018), Neanda Salvaterra reviews the report by the Energy Information Administration, which says that contrary to the effort of the climate alarmists to kill coal, coal continues to be a major source of power generation in developed and emerging nations and accounts for as much of the world's electricity today as it did in the 1990s. As it turns out, coal has proven to be incredibly resilient in Asia and Africa, where it has been pushed up by rising demand as outlined in Paul Garvey's "Fans of Coal Are Reaping the Rewards" (Garvey 2018).

In "The New Age of Coal" (Mendoza 2018), Christopher Mendoza argues that the rising demand for coal may come as a shock for anti-coal crusaders, but it should hardly be surprising when one follows the money in a free-market economy. For Mendoza, coal is cheap and readily available; that makes it an ideal fuel source for developing countries. Indeed, for some countries, exploiting their domestic coal resources is the only way they can attain economic development and create a better future for their citizens.

For Mendoza, it is incredible that climate alarmists can actively lobby to restrict funding for coal in developing countries with the naïve expectation that this effort will lead to solar panels and wind farms in countries such as those in Africa. The result is that African countries are being held back in their development. The Nigerian government has plans to generate about 30 percent of its electricity from coal. The country's Mines and Steel Ministry identified nearly three billion tons of coal reserves, which would provide the fuel crucial for the country's electricity generation, steel production, and cement manufacturing all of which have enormous economic potential. However, because of unfounded climate change ideological reasons, the World Bank and other multilateral financial institutions are denying Nigeria the necessary funding because of pollution concerns resulting in this dream being dead in the water (Abuja 2017).

Mendoza notes that the tragedy of the situation is that the arguments of the anti-coal coalition are dripping with duplicity and condescension. At a 2016 IMF meeting, former Nigerian finance minister Kemi Adeosun called out (Tijani 2016) developed countries for their "hypocritical behavior" especially since it was coal that drove Western industrialization. She's right to be angry since the technologies to achieve significant carbon dioxide emission reductions for coal and to make its use more efficient already exist. High-efficiency and low-emission technology, carbon capture, and storage technology have been designed to curb coal emissions to help countries strongly reliant on coal to strike a balance between electricity and environmental protection needs. And under Trump, the United States has emerged (Clemente 2018) as the clear leader in promoting these crucial technologies. But with die-hard activists wreaking havoc around the globe, these facts are willfully suppressed.

Keep following the money. Mendoza notes that China is planning to build as many as seven hundred coal plants in coming decades using domestic coal reserves to meet future energy demands, and developed countries are also still using coal as a primary power source. Poland has announced plans to use coal for 50 percent (Mikulska 2018) of its power generation through 2050. And Germany, the shining light for everyone with a hatred of nuclear and coal, is in fact still generating 40 percent (Board 2017) of its energy from coal.

For Mendoza, the climate alarmist narrative that the age of coal is over is blatantly wrong. Technologies such as the high-efficiency coal-fired power plants the US is developing are better exported where they are needed most rather than suppressed. Instead of blindly continuing to vilify coal while demand for energy grows all over the world, governments should spend their money and political capital investing in ways to make clean coal the standard around the globe.

Lost on the climate alarmists and the liberal big media are the economic abundances that are occurring in an environment in which Americans are fed a steady diet of dire predictions (Park 2015) of climate change with its scientifically unsupported position of human-caused global warming. Compromised scientists tell us that weather phenomena such as the extremes of storms, droughts, wind, heat, and rainfall will be more frequent and intense. Add pestilence, pollution, fires, and the encroachment of human activity to other natural calamities and one wonders just how the American farmer can survive much less prosper.

In "Will Global Warming Destroy the World? Ask America's Farmers" (Krisinger 2018), Chris J. Krisinger provides his reader with the assurance that the follow-the-money common sense is all the evidence citizens need to understand the economics going on around them.

In 2018, the USDA estimated record-setting corn and soybean crops. The corn crop (Knorr 2019) was expected to be above average for a record sixth year in a row while soybean production was projected at an all-time high of 4.4 billion bushels, up 4 percent from 2017's record. These limited crop yields confirmed once again that American farmers helped feed a hungry planet that has more than 7.6 billion inhabitants (UN 2017) whose number could reach 8.6 billion by 2030.

Keisinger notes that global agricultural trends also reflect production gains. Since 2002, world production (Looker 2016) of four major crops—corn, wheat, rice, and soybeans—has grown by 846 million tons or 48 percent. Yields have kept pace with the world's annual population growth rate of 1 percent. In fact, prices for staple grain crops reveal a downside to those abundances such that plentiful supply (Nickel 2017) depresses commodity prices on world markets. "There is too much corn," said one analyst to match demand. Corn and soybean growers now concern themselves with consumption of previous record-setting crops to promote future market price increases.

Keisinger observes that despite these blessed abundances in crop production and economic progress, Americans are forced to listen to the propaganda of climate alarmists and their scientists that weather phenomena such as the extremes of storms, droughts, wind, heat, and rainfall will be more frequent and intense.

The public does not hear that the American farmer continually adapts to the climate and weather through changes in crop rotation, planting times, genetic selection, fertilizer choices, improved equipment, innovation, pest and water management, and shifts in areas of crop production among other possible measures. Farmers take advantage of an unmatched system of education, research, science, and technology in American universities and business that has evolved to aid and support American agriculture. Farmers also make good use of a responsive agribusiness banking and finance system. On the whole, American farmers are part of and benefit from a well-honed agricultural infrastructure that fosters advances in production and efficiency.

By contract, Keisinger shifts focus to Africa (OECD 2015), which despite its vast natural resources including expanses of arable land has the world's highest incidence of undernourishment (estimated at near one in four people). It is assessed that more than 60 percent of the planet's available and unexploited cropland is in sub-Saharan Africa, but agricultural production remains dismal, which further undermines Africa's future and economic growth. Africa must import food staples valued at some $25 billion annually largely because continental food production, supply, and consumption systems do not function optimally. Keisinger argues that this poor agricultural performance is due to the fact that no nation on that continent can provide its farmers the needed political and societal stability to support a similarly developed agricultural infrastructure.

Zimbabwe (Power 2003; formerly Rhodesia and once Africa's breadbasket) is another example of the entire continent's challenges. It has Africa's most fertile farmland, but as a recent exposé explained, this "onetime net exporter of maize, cotton, beef, tobacco, roses, and sugarcane" now "exports only its educated professionals" who fled by the thousands from decades of corrupt autocratic rule. In Sudan (KPMG 2017), only 16 percent of available land had been cultivated by 2009, the majority of which now falls within South Sudan, a new country that must still import nearly all its food.

For Keisinger, it is not climate change, weather phenomena, human encroachment, or other natural calamities that pose the greatest threats to future generations. Humans adapt to their environment and can adjust their agricultural and industrial enterprises to feed the people of their own countries and those of the world. The real global threat is poor, nonfunctioning governance and more precisely autocratic, dictatorial, corrupt, socialist regimes not acting for the common good of their citizens.

19

World Population Growth and Climate Change

IN "CLIMATE Alarmists: Abort Your 'Extra' Children" (Benson 2017), Timothy Benson observed that over the past few years, outlets such as NPR, the *Atlantic Monthly*, the *Huffington Post*, and the *Guardian* have written articles with headlines such as "Should We Be Having Kids in the Age of Climate Change?" (Ludden 2016), "The Climate Change Solution No One Will Talk About" (Plautz 2014), "Voluntary Birth Control Is a Climate Change Solution Nobody Wants to Talk About" (Prois 2015), and "Though Climate Change is a Crisis, the Population Threat is Even Worse" (Emmott 2015).

Benson notes that the NPR story highlights a 2016 paper by philosophers and bioethicists at Georgetown University and Johns Hopkins University entitled "Population Engineering and the Fight against Climate Change" (Hickey 2016), which argues in favor of penalizing families for having children via a progressive tax that would increase with each child.

Posted on the Center for Biological Diversity (CBD) website is the article "Human Population Growth and Climate Change" (CBD 2017), in which Benson says,

> People around the world are beginning to address
> [climate change] by reducing their carbon footprint

> through less consumption and better technology ... But
> unsustainable human population growth can overwhelm
> those efforts, leading us to conclude that we not only
> need smaller footprints, but fewer feet ... Long-term
> population reduction to ecologically sustainable levels
> will solve the global warming crisis and move us to
> toward a healthier, more stable, post-fossil fuel, post-
> growth addicted society.

After the Census Bureau reported on May 7, 2017, that the US
population surpassed 325 million for the first time, CBD, through its
web organization Endangered Species Condoms put out a press release
(CBD 2017) stating,

> Hitting this population record highlights the danger
> of the Trump administration's attacks on reproductive
> healthcare and environmental protections. We're
> crowding out wildlife and destroying wild places
> at alarming rates, and Trump's reckless actions will
> worsen the effects of our unsustainable population
> growth, overconsumption and urban sprawl.

Benson reminds readers that the doomsday predictions about
overpopulation stretch back to Thomas Malthus's 1798 *An Essay on
the Principle of Population* (Malthus 1798). Because food production
grows arithmetically while population grows geometrically, Malthus
reasoned that the planet's burgeoning population growth would eventually outstrip food supplies leading to famine and mass starvation:
"The power of population is so superior to the power of the earth to
produce subsistence for man, that premature death must in some shape
or other visit the human race."

According to Benson, twentieth-century neo-Malthusianism was
taken up by environmentalists after World War II and gained steam after the publication of Fairfield Osborn's *Our Plundered Planet* (Osborn

1948) and William Vogt's *The Road to Survival* (Vogt 1948) in 1948 and again in 2007 (Vogt et al. 2007). Vogt wrote, "It is obvious that fifty years hence the world cannot support three billion people ... Unless population increases can be stopped, we might as well give up the struggle." Benson argues that Vogt was wrong; he states that seventy years after this forecast, earth was supporting not merely 3 billion but 7.5 billion people.

Twenty years after Osborn and Vogt, Paul Ehrlich in his 1968 best-seller *The Population Bomb* (Ehrlich 1971b) predicted mass starvation in the 1970s and 1980s due to overpopulation. "The battle to feed all of humanity is over," he wrote. "In the 1970s the world will undergo famines—hundreds of millions of people will starve to death in spite of any crash program embarked upon." Again, another fake guess as these famines and mass human die-offs never materialized.

Not to be denied, Ehrlich doubled down on his doomsday prediction in a speech before the British Institute for Biology in 1971 by restating, "By the year 2000, the United Kingdom will be simply a small group of impoverished islands, inhabited by some 70 million hungry people ... If I were a gambler, I would take even money that England will not exist in the year 2000."

The Club of Rome, in *Mankind at the Turning Point* (Mersarovic 1975), its 1974 doomsday follow-up to 1972's doomsday book *The Limits to Growth* (Meadows 1972), literally says, "The World Has Cancer and the Cancer is Man." In *Merchants of Despair* (Zubrin 2013), Robert Zubrin writes, "This idea [that] humans are cancer upon the Earth, a horde of vermin whose unconstrained aspirations and appetites are endangering the natural order ... is the core idea of antihumanism." Zubrin further states,

> Antihumanism ... is not environmentalism, though it sometimes masquerades as such. Environmentalism, properly conceived, is an effort to apply practical solutions to real environmental problems, such as air and water pollution, for the purpose of making the world a

better place for all humans to thrive in. Antihumanism, in contrast, rejects the goals of advancing the cause of mankind. Rather, it uses instances of inadvertent human damage to the environment as point of agitation to promote its fundamental thesis that human beings are pathogens whose activities need to be suppressed in order to protect a fixed ecological order with interests that stand above those of humanity.

Benson argues that these calls for population-control measures to fight climate change is at its core antihuman. Zubrin continues.

Since all human activity must perforce release [carbon dioxide], all human existence is a crime against nature. Therefore, nothing we can do is right—and so, in the name of the Higher Good, we must be constrained to do as little as possible. Thus, the global warming argument recasts the basic Malthusian line in a novel form, but with the equivalent end result. Instead of claiming that human activity must be limited because there are not enough resources, it is said that what is limited is not resources, but the right to use resources. It all amounts to the same thing: there isn't enough to go around, therefore human aspirations must be crushed.

In "Bill Nye, the Eugenics Guy" (Sobieski 2017), Daniel John Sobieski cites one of the Netflix series shows, "Bill Nye Saves the World," in which the host, Bill Nye, asked a group of panelists (Nye 2017) including a Georgetown professor, "Should we have policies that penalize people for having extra kids in the developed world?" Nye's question was rhetorical, but he did imply that there were too many people and parents at least in the developed world and that they should be punished for having "too many" children (Griswold 2017) with the following statements.

"The average Nigerian emits 0.1 metric tons of carbon annually," noted Nye's guest, Dr. Travis Rieder. "How many does the average American emit? Sixteen metric tons."

Rieder said Americans having an average of two children are "waaaay more problematic" than Nigerians having seven when it comes to preventing global warming.

"Should we have policies that penalize people for having extra kids in the developed world?" Nye asked.

"I do think we should at least consider it," Rieder said.

Nye pushed him even further.

"Well, 'at least consider it' is like, 'do it,'" he opined.

Sobieski noted that the other two guests did push back and pointed out that what Nye and Rieder were proposing came dangerously close to the eugenics policies (Ko 2016) of America's past, which ended up disproportionately targeting poor women and minorities. Sobieski agrees that these remarks are perilously close to advocating eugenics. For Sobieski, it is a short trip from deciding how many children people should be allowed to have to deciding who gets to have children at all. It bears a striking resemblance to Communist China's one-child policy, which included sterilization and forced abortion. China's one-child policy instituted by their Communist government in the late 1970s was intended to stem rising population, and it compelled couples in urban areas to have just one child and limited couples in rural areas to two children if the first child was a girl as girls were seen as having lesser value than boys in some parts of the Asian nation.

Sobieski cites an article reported in the *Investor's Business Daily*, "Biden Endorses 'One Child Policy'" (IBD 2011), in which Joe Biden, Obama's former vice president, and John Holdren, Obama's former science adviser agreed with Bill Nye that too many people were a problem that needed to be dealt with.

> This administration supports draconian actions to fight climate change. In 2007, at climate change talks in Vienna, Su Wei, a senior Chinese Foreign Ministry

official, boasted that China's one-child policy had reduced China's population at that point by some 300 million human beings, roughly equal to the US population.

Avoiding those 300 million births "means we averted 1.3 billion tons of carbon dioxide in 2005" based on average world per-capita emissions of 4.2 tons, he said.

John Holdren, the former president's top science adviser, had no quarrel with this barbarism; he considered it necessary to fight global warming and resource depletion. Even the US Constitution wouldn't stand in his way.

Hillary Clinton's idol and founder of Planned Parenthood Margaret Sanger had her particular ideas on which children should be allowed to be born. J. Kenneth Blackwell, writing in the *Washington Times* (Blackwell 2015), noted those who chant "black lives matter" obviously exclude the abortion rate of black babies that Planned Parenthood founder Margaret Sanger and the KKK could only dream of.

138,539 black babies, nearly one baby in three, were killed in the womb in 2010. According to the CDC, between 2007 and 2010, innocent black babies were victimized in nearly 36 percent of the abortion deaths in the United States, though blacks represent only 12.8 percent of the population. Some say the abortion capital of America is New York City. According to LifeSiteNews, the city's Department of Health reported that in 2012, more black babies were aborted (31,328) than born (24,758). That's 55.9 percent of black babies killed before birth. Blacks represented 42.4 percent of all abortions.

For Sobieski and for all civilized societies, this is a disturbing and tragic situation that continues unabated and is the fulfillment of Sanger's dream. As Blackwell also noted,

> According to Sanger, "Colored people are like human weeds and are to be exterminated." She opened her first abortion clinics in inner cities, and it's no accident that even today, "79 percent of Planned Parenthood's abortion facilities are located in black or minority neighborhoods."

Population control is the tool of tyrants. Whether it is to build a master race or lower carbon dioxide emissions, it is inherently evil in its methods and goals. Ironically and contrary to Nye's hypothesis, wealthier societies are healthier societies and are better for the environment.

The view expressed by those climate alarmists advocating eugenics policies based on their argument that human beings are inexorably outstripping the globe's capacity to sustain them is one of the most vivid, powerful, and enduring economic myths of the modern era.

In reality, the natural population growth on earth is slowing down. In the UN report "World Population Prospects" (UN 2015), global fertility will slow from 2.5 children per woman in 2015 to 2.0 children in 2050. Eighty-three countries accounting for 46 percent of the globe's population already had below-replacement fertility levels between 2010 and 2015, and they included Brazil, China, Germany, Japan, Russia, and the US. Forty-eight countries are also projected to see their populations decline through 2050. Of these, eleven are expected to see their populations decline by more than 15 percent including Japan and most of the countries of Eastern Europe. The UN projects that global population will level off somewhere between 9.5 billion and 13.3 billion around the year 2100 before beginning to drop.

Providing specific details on the effects of population decline around the world, Eileen F. Toplansky ("Population Stabilization or Suicidal Demographics?" Toplansky 2018) cites an article by Giulio

Meotti, "Europe demographic suicide: See Greece" (Meotti 2018a), in which he writes that it is estimated that there will be a reduction of about 25 percent in the Greek population by 2050. Even more worrying is the forecast by the country's statistical agency (Elstat), that by 2080 the population of the country could fall to 7.2 million ... Births in public hospitals have dropped by 30 percent and Greece has become a world leader in abortion ... Only Italy has a higher percentage of older people ... It is no coincidence that the three European countries considered at risk of default in the present period—Italy, Greece and Portugal—are also those most overwhelmed by the demographic winter.

Toplansky cites Peter Kotecki's "10 countries at risk of becoming demographic time bombs" (Kotecki 2018) at Business Insider; Kotecki explains,

> All around the world, countries are seeing dwindling birth rates and rising life expectancy. Aging populations are leading to greater spending on healthcare and pensions, but the number of people working and paying taxes is steadily decreasing. As a result, these countries are at risk of becoming 'demographic time bombs,' signifying a crisis of too few working people. Demographers say countries need fertility rates of 2.2 children per woman to maintain a stable population, but many nations' birth rates, such as those of South Korea and the United States, have fallen below 2.

The US birth rate has remained below replacement level since the 1970s, which means not enough children are being born to keep the population at a steady level.

Toplansky notes that in Spain, there are more deaths than births each year and that some towns are nearly abandoned. It is vital to realize that Spain (Meotti 2018b) is "a demographically dying country and a land of great investments for Saudis and other emirs"—all of whom are ultimately concerned with establishing a global caliphate—hence, the population increase in 2017 in Spain resulted from Muslim migrant arrivals.

And while "the number of immigrants to Italy rose last year and the number of Italians leaving the country decreased ... neither trend is reversing Italy's path toward a demographic time bomb."

Toplansky turns her focus to Bulgaria, whose population is shrinking faster than any other nation in the world. Bulgaria's fertility rate is only 1.46 children per woman. In 2017, Bulgaria's deputy labor minister, Sultanka Petrova, stated, "The decline in the active population is a social and economic bomb that will explode unless we take adequate measures."

Latvia's population is steadily dropping; many Latvians are leaving to look for jobs in other parts of the European Union.

Some South Koreans cite a lack of financial stability as one of the main reasons they are not having children. In order to stem the decline, the "South Korean government has offered cash incentives to people who have more than one child, as the fertility rate currently sits (CIA 2019) at 1.26 children per woman—too low to maintain a stable population."

Likewise, the Japanese government is offering cash incentives in an effort to encourage women to have children. Researchers are worried (CFPP 2018) about a demographic time bomb in Japan, where the lowest-ever number of births was recorded in 2017. Japan has

> one of the highest life expectancies in the world, leading it to face a steep decline at one end of the lifecycle and a boom at the other. Similarly, the proportion of the working age population is falling which constitutes a significant challenge for the economy.

> Economic concerns are one of the most frequently cited reasons for Japanese people to get married later in life or remain single. Women who need, or want,

> to work face difficulties in combining employment
> and child rearing, due to the limited availability of
> childcare services, unfavorable employment practices,
> and a lack of flexible working conditions.

The United Kingdom's birth rate has fallen to its lowest level in a dozen years, and China's fertility rate keeps dropping despite the government's 2016 decision to allow families to have two children instead of one (a policy that had been in place since 1979), so local authorities are taking steps (Chan 2018) to encourage more childbirth.

For Toplansky, while the importation of foreign-born migrants may show an increase in actual population numbers, it is unknown how these immigrants will affect a country's culture should they not assimilate. Toplansky asks his readers to look at Europe and see how Western values, language, and culture are being effaced.

Toplansky cites the following online posting "15 Drastic Effects of Population Decline" (GUFF 2016), which provides a list of effects on an economy by a population decline in industrial countries.

Contrary to the troubling population statistics and suggested calls for population-control measures, Toplansky cites John Seager's posting "Lower Birth Rate Trigger False Alarm" (Seager 2018) on the Population Connection website, where Seager strongly disagrees with many practitioners of the dismal science when it comes to the impact of lower birth rates. Seager asserts that "no one need worry about slower (or better yet, zero) population growth. In a world plagued with all manner of shortages (clean air, fresh water, food, common sense), we face no people shortage."

Although Toplansky cites Seager's arguments, she has real concern with the trend in the world population and the potential impact this situation can have on the progress that has already been made around the world with its citizens achieving decent standards of living.

With the viability of economic stability being threatened by population replacement levels, there is a risk that worldwide social accomplishments achieved to date may be at wisk. According to Ivana

Kottasova ("World poverty rate to fall below 10 percent for the first time," Kottasova 2015), the number of people living in "extreme poverty" ($1.90 or less per day) fell below 10 percent for the first time in human history in 2015 according to World Bank. As David Harsanyi explained (Harsanyi 2017) in *The Federalist,* not only are fewer people living in extreme poverty, but fewer are hungry than ever (Poon 2015); fewer die in conflicts (Tsang 2014) over resources, and deaths due to extreme weather have been dramatically declining for a century (Goklany 2009) ... Over the past 40 years, our water and air has become cleaner, despite a huge spike in population growth. Some of the Earth's richest people live in some of its densest cities.

Indeed, the earth is cleaner and safer today than at any point in the lifetime of anyone now living. Technology and human ingenuity have dramatically reduced the human impact on the environment. For example, by increasing crop yields (thanks, GMOs!), we are growing more food while devoting less land to agriculture. Human innovational and technological advancements have improved both the lot of humanity and the environment.

20

God and Free Markets Will Adapt to Climate Change

HAMLET'S QUESTION is not "to be or not to be" or even "to act or not to act"; it is "to believe or not to believe."

In "Being Human and the Abuse of Science" (DeBlasi 2018), Anthony J. DeBlasi opens with the statement,

> Human being calls for living according to a set of beliefs that come from the source of being—not from the brain, not from science, not from natural or man-made objects. Being human stems from cognition that transcends the opinions and calculations of any individual or group. This is something that people have always known and leaders have forgotten … Philosophers and theologians who object or rationalize around this central fact of human life ignore or make light of the fact that we do not put ourselves here and know zilch about how, for example, water, food, and air become thoughts, emotions, and the countless products of human life and civilization, from safety standards to works of art. This large blind spot regarding the reality

of the world and ourselves leads to endless falsehoods that obstruct sound judgment and action.

For DeBlasi, one great and seriously obstructing falsehood is that science liberates us from ignorance about ourselves. But if it were up to scientists like DeBlasi, they would drop dead because they wouldn't know how to manage the zillion things the body must do to keep individuals from visiting "the undiscover'd country from whose bourn no traveller returns" (Shakespeare, *Hamlet*).

DeBlasi holds that "the belief that we are 'masters of our fate' and 'captains' of our soul (Henley, W. E. "Invictus") while not pure hubris must be balanced with the realization that, after all, we are not our own gods."

DeBlasi argues that because the source of our being is obviously not science or any other system of human knowledge, science is not a legitimate basis for the beliefs that help make us human. Yet that is what it has been (ab)used for across the centuries, and the resulting bull fed the public. DeBlasi emphasizes that

> "cold facts" and empirical data by themselves are meaningless unless collected into a system of thought and action *believed* to be true. It is sometimes argued that, as a way to "make sense of the world," science is in an exclusive class by itself, but as *serious belief*, science is not exempt from competition with other serious beliefs.

For DeBlasi, science is a great and wonderful tool. But to consider it a gateway to action consistent with being human is intensely wrong minded unless the object is to turn people into machines of some kind. Not laughably, this is what some futurists (Wikipedia, transhumanism) are comfortable with.

Cutting whatever path through theology and philosophy including that concerning science ("Philosophy of Science," Stanford 2018), the final question must remain, which god do we believe?

DeBlasi cites a paper dated November 23, 1654, which was stitched into the lining of Blaise Pascal's coat and found after his death: "God of Abraham, God of Isaac, God of Jacob, *not of the philosophers and scholars.* Certainty. Certainty. Feeling. Joy. Peace." The famous French philosopher and mathematician who invented the adding machine and developed the modern theory of probability ultimately and privately conceded that the joy of certainty proceeded not from the mind of man but from God. To be released from anxiety and uncertainty is not through a process of calculation based on assumptions or derived from experience but through alignment with the Almighty. The brilliant Pascal realized as we all must that belief in God energizes the contact between us and the source of our being, which alone can steer us along the best track in life.

DeBlasi believes that Pascal's maxim echoes through the corridors of time. An echo, still loud and clear, comes from a nineteenth-century statesman.

> The postulates of science do not rest upon absolute knowledge, but are derived from sources similar to those of religious conviction. If only the data of physical researches and sensory evidence be allowed by thinking people, then we must labor forever in the agonies of doubt[.] ... They who begin in doubt may end in certainty through a higher skepticism ... not the narrow destructive skepticism of the egoist, deliberately seeking unbelief, but instead an intellectual recognition of the want of evidence. Skepticism need not destroy belief; it may serve, on the contrary, to expose the unjustifiable complacency of unbelievers. (Russell Kirk on thoughts of Arthur Balfour in *The Conservative Mind*, chapter 11, Kirk 2001).

Further, DeBlasi argues that one of the saddest consequences of the disconnect between human minds and their source has been the terrible game that democracy has become in the US. The hidden rule of the game is that everything sacred or not, negotiable or not, is subject to the tyranny of a majority, whose members believe what they are made to believe.

Cited in "Twilight of the Green Follies" (Alexiev 2018), Alex Alexiev states,

> for more than two and a half centuries, human kind has lived under an irreconcilable dichotomy—the benevolent revolution we call the enlightenment, and the inevitable reactionary counter-revolution that followed it—a dichotomy that has continued to our days.

Alexiev argues,

> The Enlightenment introduced a number of revolutionary concepts that demolished the church dogma that had dominated the Middle Ages. It established reason and empirical knowledge as the source of authority leading to the scientific revolution beginning with Copernicus and the heliocentric theory of the universe. In government, the enlightenment brought about the radical idea of individual liberty with John Locke's call for "life, liberty and property." For Alexiev, the revolution reached its apotheosis in the late 18th century with the American Constitution and its idea of "inalienable rights" given to its citizens by their Creator and of a government based on the consent of the governed. All of this was based on the unshakeable belief in progress driven by man and the Judeo-Christian civilization's fundamental belief in the primacy of man over nature.

Alexiev further adds that, no sooner did these radical ideas gain wide currency in the West than the reactionary counterassault materialized. It started with Jean-Jacque Rousseau, considered by many the father of the totalitarian temptation, and his idea of an all-powerful state using coercion as means of imposing an imaginable "general will." Since then, humanity has struggled to reconcile two ideologies that are fundamentally at odds: one based on the rights of the individual, the other espousing the unlimited power of the state. The latter one found its culmination in the bloody totalitarian ideologies of the 20th century, best expressed in Mussolini's dictum "everything within the state, nothing outside the state, nothing against the state." And it is this veneration of the coercive powers of the state that fundamentally unites Nazism, fascism, and communism despite other marginal differences.

The assault by an all-powerful state on the belief in progress that underpinned the enlightenment continued in the late 18th century with Thomas Malthus' reactionary idea of excessive population as a burden to civilization, expressed today as the bogus "population bomb" scares of Paul Ehrlich and the coercive and inhuman one-child policy in China. This assault on reason veered off into the racist eugenics theories of the first half of the 20th century and ultimately found its highest expression in the cult of nature over *homo sapiens* on which all modern green ideology is based.

First expressed in 1913 in a proto-Nazi manifesto by notorious anti-Semite, Ludwig Klage (Zubrin 2017), called "Man and Earth," this has remained the central belief system of the pioneer German Green Party since its founding in 1980 and all the green movements that have followed since. Indeed, in 1980, the Green Party reissued "Man and Nature" without any reference to its reactionary authorship.

According to Alexiev, it would be a mistake to consider this cult of nature an innocent manifestation of the earning to restore nature's presumed paradise lost. Its most consequential current incarnation—"global warming"—is a barely disguised assault on capitalism. It is for many, the last and best chance of the socialists to do away with the hated free enterprise system. Although not trying to explain their thinking behind this hatred, Alexiev notes that following the collapse of the Soviet Union, the American socialist, Robert Heilbronner (Boaz 2005), admitted that the capitalist system had proven economically superior to the socialist model, but urged his co-religionists not to despair. It was still possible to achieve socialist goals, he argued, by using the ecological movement and environmentalism.

With its bogus claims of ever greater triumphs of renewable energy, government mandates and exorbitant subsidies, the global warming scam closely resembles Stalinist Lysenkoism (Alexiev 2009) and the erstwhile Soviet propaganda of the glorious achievements of socialism.

Unfortunately, reality tends to wreak havoc with the fantasies of socialist charlatans and so it has with the putative solution to the science of global warming.

In "The Humanities, Viewed as a Crucible for Higher Educations" (Silber 1973), the late John R. Silber, former president of Boston University, made the following comment on the abuse of science.

> Scientific programs that are powerfully effective in understanding and controlling largely isolated data in such fields as physics, have been applied crudely, reductively, and disastrously in the humanities and social sciences. Discipline after discipline has succumbed to the dogma that only the quantifiable is true.

In "The Folly of Scientism" (Hughes 2012), Austin L. Hughes wrote,

> When I decided on a scientific career, one of the things that appealed to me about science was the modesty of its practitioners. The typical scientist seemed to be a person who knew one small corner of the natural world and knew it very well, better than most other

human beings living and better even than most who
had ever lived. But outside of their circumscribed
areas of expertise, scientists would hesitate to
express an authoritative opinion. This attitude was
attractive precisely because it stood in sharp contrast
to the arrogance of the philosophers of the positivist
tradition, who claimed for science and its practitioners
a broad authority with which many practicing scientists
themselves were uncomfortable.

In "Hijacked Science" (DeBlasi 2018), DeBlasi argues,

Professionals who rely on the domain of the so-called
social and political "sciences" for their base of action
are de facto pseudoscientific speculators, often at
public expense. But the speculations of vulgarized
science distill to bull, regardless of how "scientific" the
language, tight the system, convincing the supporting
data, or smart the arguments.

Further, DeBlasi states, pseudoscientists pretend to explain human
activity as though it were a lab exercise, presumably to arrive at guide-
lines for what is judged best (by them) for the conduct of society. Their
output has been called "evidence-based policy-making." *Evidence?*
Check a slogan that has been used to "justify" this approach: "Science
is the pursuit of knowledge, knowledge is power, and power is politics."

Notice the circle of reductions, linking output with input, a trick
of leftist "liberalism" to hoodwink everyone (including the "liberals" themselves). And notice the implication that science is the only
way of knowing what is best for people. So much for the value of such
"evidence."

DeBlasi asks, in today world, where is there strong competition
from less "verifiable" but substantially valuable modes of insight, such
as provided by literature and the fine arts, philosophy and theology?

Being "scientific" is not the only criterion for what is verifiable. True scientists do not look down on methods of knowing outside of their domain, aware of the very real limitations of their work. In fact, some have tapped into that vast reservoir of "unquantifiables" so despised by positivists, such as the divine inspirations that formed the complex mathematical theorems in the mind of Ramanujan (Wikipedia, Ramanujan).

According to DeBlasi, it is unfortunate that

> the bull from pseudoscientists compounds the inherent errors of the trade and magnifies the problems targeted for solution. Yet these dealers in false science expect you, him, and everybody else to live by their absurd programs, via NGOs and government agencies, using tactics of coercion that rival any in the history of oppressive rule. Some of us notice that the rule by iron hand in some autocracies has been replaced in our democratic republic by the rule of a majority brainwashed by controlled liberal big media. We also notice that the ethics employed in such counterfeit democracy have more of might than morality behind them, adjusted ad lib to the political climate.

No one will admit it, but scientists in general have become the collective witch doctor of contemporary society. They have spooked the public with a dazzling array of experiments and body of knowledge that would indicate a seemingly infinite power to influence the world, staging ever more daring feats of scientific wizardry. This has made too many to overlook the fact that the world is also profoundly influenced by nonscientists like Confucius, Plato, Christ, the originators of our musical scale ... Gandhi ... (name your own agents of progress who were not practicing scientists).

DeBlais admits that *full* knowledge of the world, macro and micro—outer and inner—perpetually recedes from our reach, fading from sight like a faint star that vanishes from view when gazed at, in order to see it more distinctly. It is not easy for a proud species like ours to see, even less accept the fact, that the world (including us) is at root unknowable in the rational sense. This should be the starting point in any serious endeavor toward progress that affects human beings. In that endeavor, *let science be science*, for its vitally important work must not be tampered with.

For those who choose science for their vocation, if you are to remain true to your calling, must seek ways to isolate themselves from the corrupting reach of politics. "Going along to get along" is not an option.

In "Science Needs a New Paradigm" (Arvay 2017), Robert Arvay argues in support of this notion that

> science and politics used to be very separate institutions. Where they did overlap, science was nonpartisan. The role of scientists was to provide objective evidence— and dispassionate, nonpolitical interpretations of that evidence. Indeed, one rarely if ever could detect the political leanings of any particular scientist. Also, science and religion used to get along, at least for the most part.

For Arvay, today, that has changed, and the results include significant dangers for society. As an example, Arvay cites the topic of climate change which has produced the myth of "settled science." Science is never settled. While we all may agree that the climate does change, there is an anti-capitalist, political agenda behind the claims of many climate alarmist scientists—that Western societies must radically reduce their standards of living to prevent climate catastrophe. Politics and ideology, not science, promote that so-called scientific view.

There is an even deeper and darker implication involving the politicization of science. To provide an example, Arvay cites a state court case, dubbed the Scopes Monkey trial [History 1925], which tested a law that forbade the teaching of Darwinian Evolution theory in public schools. On a legal maneuver, Darwinism technically lost that particular trial, but all subsequent federal court rulings since then, have upheld the theory of evolution as accepted fact. Contrary theories are essentially forbidden. Evolution is "settled science."

Arvay examine the cultural fallout from that theory in citing a quote by the late paleontologist George Gaylord Simpson: "Man is the result of a purposeless and natural process that did not have him in mind. He was not planned."

What Darwin and Simpson have done, along with others, is to introduce into society the physicalist paradigm, the one that holds that nothing exists except stuff, that is, material reality. According to physicalism, there is no spirit, no God, no eternal afterlife. By extension of this paradigm, individuals are nothing more than stuff, that is, the atoms that make up our physical bodies. If that is to be considered true, if one follows this line of thinking, individuals have no inherent right to be treated as anything more than protoplasm, nothing more than just another species of animal.

Arvay argues that these dismal interpretations of science are not at all scientific but only ideological. Most people, however, confronted with the scientific arguments for physics devoid of spirit find themselves ill-equipped to counter those arguments. All too many people have subscribed to the material paradigm and have come to regard religious faith as mere superstition at best and as harmful at worst.

The late Bishop Fulton J. Sheen wrote it well concerning scientific atheism,

> The great arcana of Divine Mysteries cannot be known by reason, but only by Revelation. Reason can however, once in possession of these truths, offer persuasions to show that they are not only not contrary to reason, or destructive of nature, but eminently suited to a scientific temper of mind and the perfection of all that is best in human nature. (The Life of All Living, Sheen 1979)

By contrast, the God paradigm holds that life is not merely a chemical reaction. It informs us that our free will empowers us—supernaturally—to break the otherwise immutable chain of cause and effect. This paradigm finds its meaning not only in the Bible but also in the US Declaration of Independence, which states that we are endowed by our Creator with certain inalienable rights, including life and liberty.

Arvay writes,

> This is a critical central tenet of our modern civilization. It informs citizens that their rights do not come from the government, but from God. No government has the right to infringe on those rights. Government is not the ultimate moral authority. It must be constrained to its limited functions.

The physicalist paradigm is not only morally wrong, it is also unscientific. The universe itself provides overwhelming evidence of planning and purpose by an Intelligent Designer (Creator, God). The proof is so complete that, in order to refute it, scientists have had to resort to a thoroughly unsupported, unscientific speculation that there are an infinite number of universes, whereby one of them was destined to be, purely by chance, like ours.

Although the notion of an infinite number of universes is extreme, Arvay argues,

> When scientists resort to fantasy instead of observable, repeatable experiments by skeptics, then we have abandoned reason, and begun slouching toward barbarism.

There is more than enough evidence for science to reject the physicalist paradigm, and to move toward a God paradigm. While neither can be absolutely proved by the rules of science, the world view adopted by scientists has enormous power to direct the efforts of science.

To this point, Arvay argues that science has bestowed enormous benefits on humankind while exhibiting a dark side as well. It gives us miracle medicines but also germ warfare. It bestows upon us nuclear power and nuclear bombs. Its power can be used to benefit the environment or to destroy it.

To present a "glass half empty" perspective of science, Anthony J. Sadar ("Reflections on the State of Climate Science," Sadar 2014), argues,

> For the most part, graduates in fields related to the environment (such as environmental engineering and science, biology, meteorology, ecology, and the like) have been well trained in the fundamentals of each discipline, yet at the same time have been somewhat indoctrinated in a perspective that is "progressive" ... socialism.

This progressive groupthink, which is evangelized by academic, government, and business progressives, convolutes some of the principles and language and techniques of free markets to promote their pseudoscience, programs that would be hard-pressed to survive without the assistance of free markets. Enticing concepts of entrepreneurship and profiting from energy efficiency are proffered as if they are the proprietorship of progressivism.

Furthermore, apart from the apparent indoctrination from grade school through graduate school that has inculcated the "incontrovertible conclusion" that people are ruthlessly destroying the planet, man-made climate change hype via the media and propaganda spin doctors has infused acceptance of boundless human culpability into the psyche of everyone from the general public to atmospheric-science practitioners.

Sadar notes,

> Along with the continued onerous, complex federal regulations, there are literally billions of dollars

available for researchers securing grant money, consultants advising on carbon credits, and technocrats proposing carbon dioxide control and sequestration contraptions.

From Sadar's perspective, it is easy to cash in on assessing "the risk of human-induced climate change, its potential impacts and options for adaptation and mitigation" as stated in the role of the UN's IPCC.

Arguing the "glass half full" perspective, H. Sterling Burnett ("Hooray for Carbon Dioxide! It's Helping to Feed the World's Hungry," (Burnett 2017b), begins by stating, among the greatest challenge humankind has faced throughout its history, feeding the world's hungry ranks at or near the very top of the list. And with the world's population expected to top at between 9.5 billion and 13.3 billion around the year 2100, this issue will surely become even more important in the coming decades.

The carbon dioxide humans have been pumping into the air since the middle of the 20th century has enriched plant growth and has scientifically been proven to contribute to record crop yields, which has helped to bring about the largest decline in hunger, starvation, and malnutrition in human history.

Burnett cites lessons from history that support the fact the world's plant life arose during times when carbon dioxide levels were much higher than they are today.

Over time, the amount of carbon dioxide in the atmosphere slowly declined, to the extent that during the most recent ice age, atmospheric carbon dioxide levels fell to dangerously low levels—just 180 parts per million (ppm). Plants begin to die when carbon dioxide reaches 150 ppm, because they are unable to use sunlight to photosynthesize food from carbon dioxide and water. After humans emerged from the previous ice age, carbon dioxide levels rose to approximately 280 ppm, still far below the levels existing when plant life began to colonize the land.

If plants die, humans and almost all other living beings on Earth will perish as well. Said differently, the higher the carbon dioxide levels in the atmosphere, the more life that can be supported on Earth.

Since the widespread development and use of fossil fuels, world poverty and hunger have declined precipitously. Despite adding 3.2 billion people to the planet since 1968, poverty and hunger have fallen at a faster rate than at any time in human history.

Contrary to the predictions made by 1968 Malthusian environmentalists such as Paul Ehrlich, who said in his woefully mistaken 1968 jeremiad *The Population Bomb*, "The battle to feed all of humanity is over. In the 1970s hundreds of millions of people will starve to death in spite of any crash programs embarked upon now," more people are better fed today than ever before.

Forty-four percent of the world's population lived in absolute poverty in 1981. Since then, the share of people living in extreme poverty fell below 10 percent in 2015. And although 700 million people worldwide still suffer from persistent hunger, according to the UN, hunger has declined by two billion people since 1990. Additionally, research shows there is now 17 percent more food available per person than there was 30 years ago.

This food abundance arose even as the amount of land devoted to agriculture declined over the same period, with former farm fields reclaimed by forests and pastures.

Complementing these "glass half full" conclusions, in the cogent 2007 published research paper by Robinson, Robinson, and Soon, "Environmental Effects of Increased Atmospheric Carbon Dioxide" (Robinson et al. 2007), the authors provide empirical evidence that invalidates climate alarmists' hypotheses. They also found overwhelming support for the general benefits that are derived from natural global warming.

The following is the summary of their findings.

1. A review of the research literature concerning the environmental consequences of increased levels of atmospheric carbon dioxide leads to the conclusion that increases during the 20th and

early 21st centuries have produced no deleterious effects upon Earth's weather and climate. There are no experimental data to support the hypothesis that increases in human hydrocarbon use or in atmospheric carbon dioxide and other greenhouse gases are causing or can be expected to cause unfavorable changes in global temperatures, weather, or landscape. There is no reason to limit human production of CO_2, CH4, and other minor greenhouse gases as has been proposed.

2. Predictions of catastrophic global warming are based on computer climate modeling, a branch of science still in its infancy. The empirical evidence - actual measurements of Earth's temperature and climate - shows no man-made warming trend. Indeed, during four of the seven decades since 1940 when average carbon dioxide levels steadily increased, US average temperatures were actually decreasing.

3. Increased carbon dioxide has, however, markedly increased plant growth. Predictions of harmful climatic effects due to future increases in hydrocarbon use and minor greenhouse gases like carbon dioxide do not conform to current experimental knowledge.

4. While major greenhouse gas H2O substantially warms the Earth, minor greenhouse gases such as carbon dioxide have little effect. ... The 6-fold increase in hydrocarbon use since 1940 has had no noticeable effect on atmospheric temperature or on the trend in glacier length.

5. Solar activity and US surface temperatures are closely correlated ..., but US surface temperature and world hydrocarbon use are not correlated.

6. We also need not worry about environmental calamities even if the current natural warming trend continues. The Earth has been much warmer during the past 3,000 years without catastrophic effects. Warmer weather extends growing seasons and generally improves the habitability of colder regions.

7. Human use of coal, oil, and natural gas has not measurably warmed the Earth, and the extrapolation of current trends shows that it will not do so in the foreseeable future. The carbon dioxide produced does, however, accelerate the growth rates of plants and also permits plants to grow in drier regions. Animal life, which depends upon plants, also flourishes, and the diversity of plant and animal life is increased.

With persistent broadminded scientific practice and the continued unfolding of climate conditions in ways not predicted by vaunted climate models, the future looks warm indeed for a turnaround in climate science thinking. Burnett states (Burnett 2017),

> Humans' carbon dioxide emissions have greened Earth, transforming some former desert regions into verdant oases of greenery, and contributed to record crop yields.

Contrary to the many dire predictions made by Ehrlich in 1968 and others since, humans have moved much closer to the truth captured by a *New York Times* headline from May 2016: "Is the Era of Great Famines Over?" The answer appears to be yes. Political decisions and wars, not food scarcity, is now usually responsible when populations face starvation or malnutrition.

We should praise carbon dioxide for helping to feed the world, not demonize it. Carbon dioxide is not a pollutant as so many erroneously or misleadingly suggest; it's entirely natural and vital to all life on Earth.

References

Abdussamatov, Habibullo. 2011. "Top Russian Scientist Predicts 100 Years of Global Cooling." InfiniteInknowen.net. https://infiniteunknown. net/2011/11/11/top-russian-scientist-predicts-100-years-of-global-cooling/. November 11, 2011.

Abuja, Okechukwu Nnodim. "W'Bank, others may not fund Nigeria's coal-to-power projects – Report." Punchng. https://punchng.com/wbank-other s-may-not-fund-nigerias-coal-to-power-projects-report/. August 22, 2017.

Adams, David. 2007. "Gore's climate film has scientific errors—judge." *Guardian.* https://www.theguardian.com/environment/2007/oct/11/climatechan-gel. October 15, 2007.

Agrawala, Shardul. 1997. "Explaining the Evolution of the IPCC Structure and Process." Belfercenter. https://www.belfercenter.org/publication/explainin g-evolution-ipcc-structure-and-process. July 31 1997.

Ahlseen, Mark. 2018. "The Cost of the United Kingdom's Energy Policy." American Thinker. https://www.americanthinker.com/articles/2018/04/ the_cost_of_the_united_kingdoms_energy_policy.html. April 30, 2018.

Aiken, Ann. 2015. *Juliana v. United States Youth Climate Lawsuit.* Our Children's Trust. https://www.ourchildrenstrust.org/juliana-v-us. 2015.

Alexiev, Alex. 2009. "Comrade Lysenko in Copenhagen." National Review. https://www.nationalreview.com/2009/12/comrade-lysenko-copenhage n-alex-alexiev/. December 8, 2009.

—————. 2018. "Twilight of the Green Follies." American Thinker. https:// www.americanthinker.com/articles/2018/11/twilight_of_the_green_fol-lies.html. November 2, 2018.

Allegre, Claude. 2012. "No Need to Panic about Global Warming." SyteReitz/*Washington Post.* https://sytereitz.com/2012/01/no-need-to-panic-about-global-warming/. January 27, 2012.

Allen, Greg. 2017. "South Florida Real Estate Boom Not Dampened By Sea Level Rise." Npr.org. https://www.npr.org/2017/12/05/567264841/south-florid a-real-estate-boom-not-dampened-by-sea-level-rise. December 5, 2017.

Amednews. 2011. "Confronting health issues of climate change." amednews. https://amednews.com/article/20110404/opinion/304049959/4/. April 4, 2011.

AMS. 2019. The Thinking Person's Guide to Climate Change. American Meteorological Society. https://bookstore.ametsoc.org/catalog/book/ thinking-persons-guide-climate-change-2-ed. 2019.

Anderegg, William R. L., and James W. Prall. 2010. "Expert Credentials in Climate Change," Climateviews. http://www.climateviews.com/up-loads/6/0/1/0/60100361/ posteragu2010dist.pdf. April 9, 2010.

AOML. NOAA. 2018. "What is the complete list of continental U.S. landfalling hurricanes?" AOMI.NOAA.gov. https://www.aoml.noaa.gov/hrd/tcfaq/ E23.html.

AP. 2005. "Kerry got slow start at Yale, transcript shows." NBC News. http:// www.nbcnews.com/id/8127403/ns/politics/t/kerry-got-slow-start-yal e-transcript-shows/#.XKDjfPZFxPZ. June 7, 2005.

APC. 1992. "Heidelberg Appeal," American Policy Center. https://americanpol-icy.org/2002/03/29/the-heidelberg-appeal/. March 29, 1992.

Applegate, Zoe. 2013. "Guy Stewart Callendar: Global warming discovery marked." BBC News. 37. April 26, 2013.

Archibald, David. 2018. "Someone Send the Coal People the Memo" American Thinker. https://www.americanthinker.com/articles/2018/05/someone_ send_the_coal_people_the_memo.html. May 3, 2018.

Arvay, Robert. 2017. "Science Needs a New Paradigm." American Thinker. https://www.americanthinker.com/articles/2017/11/science_needs_a_ new_paradigm.html. November 27, 2017.

Aspen. 2017. "Aspen Mountain will open for skiing Memorial Day weekend." Denver Post. https://www.denverpost.com/2017/05/22/aspen-mountai n-memorial-day-weekend-open/. May 22, 2017.

Ausubel, Jesse H. 2004. "Big Green Energy Machines." Phe.rockerfeller.edu. https://phe.rockefeller.edu/docs/BigGreen.pdf. September 21, 2004.

Avery, Dennis. 2017. "Democrats' Real Global Warming Fraud Revealed." American Thinker. https://www.americanthinker.com/articles/2017/02/ democrats_real_global_warming_fraud_revealed.html. February 18, 2017.

Ball, Tim. 2015. "Thanks To The IPCC, the Public Doesn't Know Water Vapor Is Most Important Greenhouse Gas." WattsupWithThat. https://wattsupwith-that.com/2015/02/08/thanks-to-the-ipcc-the-public-doesnt-know-water-v apor-is-most-important-greenhouse-gas/. February 8, 2015.

Barkoukis, Leah. 2016. "Democratic Party Platform Calls For Prosecuting Climate Change Skeptics." Townhall. https://townhall.com/tipsheet/leah-barkoukis/2016/06/28/ democratic-party-platform-calls-for-investigation s-of-global-warming-skeptics-n2183997. June 28, 2016.

Bast, Joseph, and Roy Spencer. 2014. "The Myth of the Climate Change '97%'." Blog.heartland.org. http://blog.heartland.org/2014/06/the-myth-of-th e-climate-change-97/. June 30, 2014.

Bastasch, Michael. 2015. "25 years of predicting the global warming 'tipping point'." *Daily Caller.* https://dailycaller.com/2015/05/04/25-years-of-pre-dicting-the-global-warming-tipping-point/. April 5, 2015.

Bates, John. 2017. "Climate scientists versus climate data." Judith Curry. https://judithcurry.com/2017/02/04/climate-scientists-versus-climate-data/#more-22794. February 4, 2017.

Battig, Charles. 2014. "Climate Change Hysteria and the Madness of Crowds." American Thinker. https://www.americanthinker.com/articles/2014/07/climate_change_hysteria_and_the_madness_of_crowds.html. July 7, 2014.

—————. 2015. "Global Warming and Government Work." American Thinker. https://www.americanthinker.com/articles/2015/01/global_warming_and_government_work.html. January 24, 2015.

Beale, Lauren. 2010. "Al Gore Buys $8.9 Million Ocean-view Villa." Worldpropertyjournal. http://www.worldpropertyjournal.com/featured-columnists/celebrity-homes-column-al-gore-tipper-gore-oprah-w infrey-michael-douglas-christopher-lloyd-fred-couples-nicolas-cage-peter-reckell-kelly-moneymaker-2525.php. May 13, 2010.

Benson, Timothy. 2017. "Climate Alarmists: Abort Your 'Extra' Children." American Thinker. https://www.americanthinker.com/articles/2017/06/climate_alarmists_abort_your_extra_children.html. June 10, 2017.

Biello, David. 2006. "Mysterious Stabilization of Atmospheric Methane May Buy Time in Race to Stop Global Warming." *Geophysical Research Letters.* https://www.scientificamerican.com/article/mysterious-stabilization/. November 23, 2006.

Birdnow, Timothy. 2011. "Warmist Cargo Cult Science Returns."

American Thinker. https://www.americanthinker.com/articles/2011/06/warmist_cargo_cult_science_returns.html. June 28, 2011.

Blackwell, J. Kenneth. 2015. "Aborting Black America." *Washington Times*. https://www.washingtontimes.com/news/2015/jan/21/j-kenneth-blackwell-black-abortions-a-crisis-in-am/. January 21, 2015.

Boaz, David. 2005. "The Man Who Told The Truth." Reason. https://reason.com/archives/2005/01/21/the-man-who-told-the-truth. January 21, 2005.

Bolin, Bert. 2008. "Bert Bolin: Meteorologist and first chair of the IPCC who cajoled the world into action on climate change" Independent. https://www.independent.co.uk/news/obituaries/bert-bolin-meteorologist-and-first-chair-of-the-ipcc-who-cajoled-the-world-into-action-on-climate-768355.html. January 5, 2008.

Booker, Christopher. 2018. Global Warming—A case study in groupthink. Thegwph.org. https://www.thegwpf.org/content/uploads/2018/02/Groupthink.pdf. GWPF Report 28, 2018.

Broeker, Wallace, et al. 1999. Hudson Canyon Map: Earth Observatory, Columbia Univ. Lecture at Amer Geographic Society, Baltimore, Spring 1999. http://maps.thefullwiki.org/Hudson_Canyon. Current.

Broeker, Wallace. 1999. Earth Observatory. Hudson Canyon Map. Columbia Univ. Lecture at Amer Geographic Society, Baltimore, Spring 1999.

Burnett, H. Sterling. 2017a. "The Carbon Tax Rebate Scam." American Thinker. https://www.americanthinker.com/articles/2017/05/the_carbontax_rebate_scam.html. May 23, 2017.

—————. 2017b. "Hooray for Carbon Dioxide! It's Helping to Feed the World's Hungry." American Thinker. https://www.americanthinker.com/articles/2017/09/hooray_for_carbon_dioxide_its_helping_to_feed_the_worlds_hungry.html. September 0, 2017.

Burnett, H. Sterling, and Justin Haskins. 2018. "Energy Socialism Kills." American Thinker. https://www.americanthinker.com/articles/2018/08/energy_socialism_kills.html. August 25, 2018.

C&EN 2005 C&EN, August 8, 2005, 16.

Carson, Rachel. 1962. *Silent Spring*. Houghton Mifflin, Anniversary Edition. http://rachelcarson.org/SilentSpring.aspx. September 27, 1962.

Cartlidge, Edwin. 2013. "Physicists claim further evidence of link between cosmic rays and cloud formation." direct relationship." Physics World. https://physicsworld.com/a/physicists-claim-further-evidence-of-link-between-cosmic-rays-and-cloud-formation/. September 9, 2013.

Cashill, Jack. 2016. "Trump, the Times, and the Coming Eco-Apocalypse." American Thinker. https://www.americanthinker.com/articles/2016/11/ trump_the_times_and_the_coming_ecoapocalypse.html. November 28, 2016.

CBD. 2017. "Human Population Growth and Climate Change," biologicaldiversity.org. https://www.biologicaldiversity.org/programs/population_and_ sustainability/climate/. 2017.

CBD. 2017. "U.S. Population Reaches New Milestone." Ecowatch. https://www. ecowatch.com/us-population-census-bureau-2398075424.html. 2017.

CConcerns, Climate. 2014. "Interglacial Comparisons." Climate Concerns. https://oz4caster.wordpress.com/2014/11/24/interglacial-comparison/. November 24, 2014.

CDAC. 1983. "Carbon dioxide Assessment Committee." National Academy Press. https://www.nap.edu/read/18714/chapter/1. October 31, 1983.

CDepot. 2010. "More Than 1000 International Scientists Dissent Over Man-Made Global Warming Claims." Amherst.edu Climate Depot. https://www. amherst.edu/media/view/400467/original/2010_Senate_Minority_ Report.pdf. December 8, 2010.

CFPP. 2018. "Tackling the Declining Birth Rate in Japan." Centreforpublicimpact. org. https://www.centreforpublicimpact.org/case-study/tackling-declinin g-birth-rate-japan/. 2018.

Chamberland, Dennis. 2015. "The Tyranny of Consensus." American Thinker. https://www.americanthinker.com/articles/2015/09/the_tyranny_of_ consensus_.html. September 2, 2015.

Chan, Tara Francis. 2018. "Chinese authorities are offering wedding subsidies and cash payments to lure 'high quality' women into having more babies." Business Insider. https://www.businessinsider.com/china-one-child-polic y-implications-women-children-2018-7. July 30, 2018.

Chow, Dennis. 2011. "How the Sun's 11-Year Solar Cycle Works." Livescience. https://www.livescience.com/33345-solar-cycle-sun-activity.html. June 15, 2011.

CIA. 2019. "The World FactBook." CIA.gov/library. https://www.cia.gov/library/publications/the-world-factbook/geos/ks.html. 2019.

Clancy, T. R. 2018. "'Future Generations' Sue the USA over Global Warming." Competitive Enterprise Institute. https://cei.org/blog/ oleary-casts-bleary-eye-eco-nutters. September 17, 2007.

Clemente, Jude. 2018. "The United States As A Clean Coal Leader." *Forbes*. https://www.forbes.com/sites/judeclemente/2018/02/14/the-united-state s-as-a-clean-coal-leader/#7b680aa21c38. February 14, 2018.

ClimateConcern. 2014. Interglacial Comparisons. OZ4caster.wordpress. https://oz4caster.wordpress.com/2014/11/24/interglacial-comparison/. November 24, 2014.

Climatecost. 2011. "Statement of European Climate Scientists on Actions to Protect Global Climate." Climatecost.cc. http://www.climatecost.cc/im-ages/Policy_brief_1_Projections_05_lowres.pdf. 2011.

Cockburn, Harry. 2017. "Global warming may be occurring more slowly than previously thought, study suggests." Independent. https://www.inde-pendent.co.uk/environment/climate-change-global-warming-paris-clim ate-agreement-nature-geoscience-myles-allen-michael-grubb-a7954496. html. September 19, 2017.

Concha, Joe. 2017. "O'Keefe video shows CNN's Van Jones calling Russia story a 'nothingburger'." The Hill. https://thehill.com/homenews/media/339867-okeefe-video-shows-cnns-van-jones-calling-russia-story-a-nothingburger. June 28, 2017.

Congress. 2006. "Questions Surrounding the 'Hockey Stick' Temperature Studies: Implications for Climate Change Assessments." Committee on Energy and Commerce—House of Representatives. https://www.govinfo. gov/content/pkg/CHRG-109hhrg31362/pdf/CHRG-109hhrg31362.pdf. July 27, 2006.

Cook, Russell. 2013. "Jumping the Shark on Global Warming." American Thinker. https://www.americanthinker.com/blog/2013/08/jumping_ the_shark_moment_on_global_warming.html. August 30, 2013.

—————. 2016. "Gore's RICO-style prosecution of Global Warming Skeptics." American Thinker. https://www.americanthinker.com/arti-cles/2016/04/gores_ricostyle_prosecution_of_global_warming_skep-tics.html. April 15, 2016.

Coolearth. 2018. "IPCC Global Warming Special Report 2018 | What does it actually mean?" Coolearth.org. https://www.coolearth.org/2018/10/ ipcc-report-2/. October 8, 2018.

Cregersen, Erik. 2018. "Dalton Minimum." Britannica. https://www.britannica. com/science/Dalton-minimum. 2018.

Creveld, Martin van. 1899. *Die Zukunft des Krieges*. Gerling-Akad.-Verl. https:// www.amazon.com/Die-Zukunft-des-Krieges/dp/3932425049. 1899.

Crichton, Michael. 2003. "Environmentalism as a Religion". http://www. sullivan-county.com/immigration/e2.html. September 15, 2003.

Curry, Judith. 2012. "Gleick's 'integrity'." Judith Curry. https://judithcurry. com/2012/02/21/gleicks-integrity/. February 21, 2012.

—————. 2018. Judith Curry. Curry.eas.gatech.edu. http://curry.eas.gatech. edu/. 2018.

Curwood, Steve. 1995. "Global Warming." Gelbspanfiles. http://gelbspanfiles. com/wp-content/uploads/2014/03/NPR-LOE-Gelb-reposit-12-15-95-1024x591.jpg. December 15, 1995.

D'Aleo, Joseph. 2010. "Climategate: Leaked E-mails Inspired Data Analyses Show Claimed Warming Greatly Exaggerated and NOAA not CRU is Ground Zero." Icecap.us. http://icecap.us/images/uploads/ NOAAroleinclimategate.pdf. 2010.

D'Aleo, Joseph, and Anthony Watts. 2010. "Corruption of ground-based temperature records used by the UN IPCC." Climate Conscious. http:// www.climate.conscious.com.au/ documents/additional%20material/ Corruption%20of%20temperature%20data.pdf. October 2010.

Darwin, Charles. 1869. "Darwin and Natural Selection." palomar.edu. https:// www2.palomar.edu/anthro/evolve/evolve_2.htm. 1869.

Davenport, Carol, and Kendra Pierre-Louis. 2018. "U.S. Climate Report Warns of Damaged Environment and Shrinking Economy." *New York Times.* https://www.nytimes.com/2018/11/23/climate/us-climate-report.html. November 23, 2018.

Davis-Wheeler, Clare. 2000. "Louisiana Coastal Land Loss." The Louisiana Environment. http://www.tulane.edu/~bfleury/envirobio/enviroweb/ LandLoss/LandLoss.htm. 2000.

DeBlasi, Anthony J. 2018. "Being Human and the Abuse of Science." American Thinker. https://www.americanthinker.com/articles/2018/10/being_human_and_the_abuse_of_science.html. October 19, 2018.

—————. 2018. "Hijacked Science." American Thinker. https://www.americanthinker.com/articles/2018/11/hijacked_science.html. November 14, 2018.

Defyccc, et al. 2001. "Tobacco Precedent Background." Defyccc.org. https:// defyccc.com/tobacco-precedent-background/. 2001.

—————. 2018. "Defunding of Climate Realists." Defyccc.org. https://defyccc.com/defunding-climate-realists/. 2018.

Delingpole, James. 2017. "Delingpole: 'Nearly All' Recent Global Warming is Fabricated, Sturfu Finds." Breitbart. https://www.breitbart.com/

politics/2017/07/09/delingpole-nearly-all-recent-global-warming-is-fabricated-study-finds/. July 9, 2017.

DiLallo, Matt. 2014. "You can't have solar without silver." *USA Today*. https://www.usatoday.com/story/money/markets/2014/08/29/no-silver-no-solar/14756397/. August 29, 2014.

Dinan, Stephen. 2018. "Feds ask judge to overturn 'Flores' ruling; Trump blames ruling for family separations." *Washington Times*. https://www.washingtontimes.com/news/2018/jun/21/feds-ask-judge-overturn-flores-ruling/. June 21, 2018.

Dobbs, Jeff. 2007. "The two Americas of John Edwards." American Thinker. https://www.americanthinker.com/blog/2007/04/the_two_americas_of_john_edwar.html. April 19, 2007.

Doman, Linda. 2016. "EIA projects 48 percent increase in world energy consumption by 2040." U.S. Energy Information Administration. https://www.eia.gov/todayinenergy/detail.php?id=26212. May 12, 2016.

Dorian, Paul. 2015. "8:45 AM | The sun is now virtually blank during the weakest solar cycle in more than a century." Perpecta Weather. https://www.perspectaweather.com/blog/ 2015/04/30/845-am-the-sun-is-now-virtually-blank-during-the-weakest-solar-cycle-in-more-than-a-century. April 30, 2015.

Dulles, John Foster. 1998. *John Foster Dulles*. Rowman & Littlefield, Lanham MD. https://www.amazon.com/John-Foster-Dulles-Pragmatism-Biographies-ebook/dp/B00E9924FM/ref=sr_1_2?ie=UTF8&qid=1444195727&sr=8-2&keywords=john+foster+dulles&pldnSite=1. November 1, 1998.

Dumbenergy, 2018. "Cost of wind, Solar and Natural Gas Electricity at the Plant Fence." Dumbenergy. https://www.dumbenergy.com/cost-of-electricity.html. 2018.

Dunn, J. R. 2007. "Who's afraid of Global Warming?" American Thinker. https://www.americanthinker.com/articles/2007/02/whos_afraid_of_global_warming.html. February 16, 2007.

Dunn, John, and Joseph Bast. 2018. "The IPCC is still wrong on climate change. Scientist prove it." American Thinker. https://www.americanthinker.com/articles/2018/ 10/the_ipcc_is_still_wrong_on_climate_change_scientists_prove_it.html. October 8, 2018.

Dunn, John Dale. 2017. "Medical Journals and the Global Warming Noble Lie." Climate Change Dispatch. https://climatechangedispatch.com/medical-journals-and-the-global-warming-noble-lie/. November 7, 2017.

Easterbrook, Don. 2010. "The Looming Threat of Global Cooling." Western Washington University. http://myweb.wwu.edu/dbunny/pdfs/looming-threat-of-global-cooling.pdf. 2010.

Editorial Board. 2017. "Germany Is Burning Too Much Coal." Bloomberg. https://www.bloomberg.com/opinion/articles/2017-11-14/germany-is-burning-too-much-coal. November 14, 2017.

Edwards, J. Gordon. 1992. "The Lies of Rachel Carson." 21st Century Science & Technology Magazine. http://21sci-tech.com/articles/summ02/Carson.html. Summer 1992.

Edwards, Paul. *A Vast Machine: Computer Models, Climate Data, and the Politics of Global Warming.* MIT Press, Cambridge MA. https://www.amazon.com/gp/product/0262518635/ref=dbs_a_def_rwt_hsch_vapi_taft_p1_i0. February 8, 2013.

Ehrlich, Paul. 1971a. *The Population Bomb.* Ballantine Books, New York NY. https://books.google.ca/books/about/The_population_bomb.html?id=YixeAAAAIAAJ&redir_esc=y. 1971.

—————. 1971b. "Population Bomb." Faculty.Washington.edu. http://faculty.washington.edu/stevehar/Ehrlich.pdf. August 28, 1971.

—————. 1968. *The Population Bomb.* Sierra Club. Ballantine Books. https://www.amazon.ca/dp/B071RXJ697?ref_=k4w_embed_details_rh&tag=bing08-20&linkCode=kpp. 1968.

EIA. 2013. "Levelized Cost of Electricity and Levelized Avoided Cost of Electricity Methodology Supplement." US Energy Information Administration. https://www.eia.gov/renewable/workshop/gencosts/pdf/methodology_supplement.pdf. July 2013.

—————. 2019. EIA Annual Energy Outlook 2019. US Energy Information Administration. https://www.eia.gov/outlooks/aeo/data/browser/. October 2018.

Eisenhower, Dwight D. 1961. "Eisenhower's Farewell Address to the Nation." McAdams.Posc.Mu.edu. http://mcadams.posc.mu.edu/ike.htm. January 17, 1961.

Ellison, Robert. 2018. "Destroy Capitalism, Save the Climate." American Thinker. https://www.americanthinker.com/articles/2015/05/destroy_capitalism_save_the_climate.html. May 28, 2018.

Emmott, Stephen. 2015. "Though Climate Change is a Crisis, the Population Threat is Even Worse." TheGuardian.com. December 4, 2015.

EPA. 2016. "Climate Change Indicators: Sea Level" EPA.gov. https://www.epa.gov/climate-indicators/climate-change-indicators-sea-level. August 2016.

Eschenbach, Willis. 1990. "Further Problems with Kemp and Mann." Wattsupwiththat. June 26, 2011.

Essex, Christopher, Ross McKitrick, and Bjarne Anderson. 2006. "Does a Global Temperature Exist?" Uoguelph.ca. http://www.uoguelph.ca/~rmckitri/research/globaltemp/ GlobTemp.JNET.pdf. June 2006.

Evens, David. 2011. "Summary Speech on Global Warming, 2011." Science Speaks http://sciencespeak.com/rally.pdf. May 30, 2011.

Feldman, Clarice. 2007. "Á Modest Proposal to Eco-Celebs." American Thinker. https://www.americanthinker.com/blog/2007/02/a_modest_proposal_to_ecocelebs.html. February 27, 2007.

Fenig, Ethel C. 2007. "Two houses symbolize two Americas." American Thinker. https://www.americanthinker.com/blog/2007/04/two_houses_symbolize_two_ameri.html. April 19, 2007.

Ferretti, D. F., J. B. Miller, et al. 2005. "Unexpected Changes to the Global Methane Budget over the Past 2000 Years." Science, Vol. 309. http://science.sciencemag.org/content/309/5741/1714.full. September 9, 2005.

Fetzer, James H., and J. R. Dunn. 2007. "It's official: San Diego paper a propaganda organ." American Thinker. https://www.americanthinker.com/blog/2007/02/its_official_san_diego_paper_a.html. February 21, 2007.

Feynman, Richard. 2015. "Mind-Blowing Temperature Fraud At NOAA." Real Science. http://armstrongeconomics-wp.s3.amazonaws.com/2015/12/Mind-Blowing-Temperature-Fraud-At-NOAA-_-Real-Science.pdf. July 27, 2015.

Finamore, Barbara. 2010. "China Officially Associates with the Copenhagen Accord." NRDC.org. https://www.nrdc.org/experts/barbara-finamore/china-officially-associates-copenhagen-accord. March 11, 2010.

Fleshler, David. 2018. "The world has never seen a Category 6 hurricane. But the day may be coming." *Los Angeles Times*. https://www.latimes.com/nation/la-na-hurricane-strenth-20180707-story.html. July 7, 2018.

Folks, Jeffrey. 2013. "IPCC Now '95 percent Sure of Global Warming." American Thinker. https://www.americanthinker.com/articles/2013/10/ipcc_now_95_sure_of_global_warming.html. October 2, 2013.

Folks, Jeffrey. 2011. "Sustainable Nonsense." American Thinker. https://www.americanthinker.com/articles/2011/06/sustainable_nonsense.html. June 1, 2011.

Fox, Porter. 2014. "The end of snow?" *New York Times*. https://www.nytimes.com/2014/02/08/opinion/sunday/the-end-of-snow.html. February 7, 2014.

French, David. 2016. "Apocalypse Delayed." *National Review.* https://www.nationalreview.com/2016/01/al-gore-doomsday-clock-expires-climate-change-fanatics-wrong-again/. January 27, 2016.

Garvey, Paul. 2018. "Fans of Coal Are Reaping the Rewards." *Wall Street Journal.* https://www.wsj.com/articles/fans-of-coal-are-reaping-the-rewards-1537189201. September 17, 2018.

Gasparrini, Antonio, et al. 2015. "Mortality risk attributable to high and low ambient temperature: a multicountry observational study." Junksxience. https://junkscience.com/wp-content/uploads/2017/11/Gasparini-on-hot-cold-planet-death.pdf. July 25, 2015.

Gelbspan, Ross. 1997. "The Heat Is On: The Climate Crisis, The Cover-up, The Prescription." Basic Books. https://www.amazon.com/Heat-Climate-Crisis-Cover-up-Prescription/dp/0738200255. September 22, 1998.

—————. 2013. "When is a 'Pulitzer Winner' not a Pulitzer Winner?" Gelbspanfiles. http://gelbspanfiles.com/?p=434. June 19, 2013.

Gergen, David. "The Argument Culture." Pbs.org. http://w3.salemstate.edu/~jaske/courses/readings/The_Argument_Culture_An_Interview_with_Deborah_Tannen.htm. March 27, 1998.

Gillis, Justin, and Clifford Krauss. 2015. "Exxon Mobil Investigated for Possible Climate Change Lies by New York Attorney General." Governors' Wind & Solar Energy Coalition. https://governorswindenergycoalition.org/exxon-mobil-investigated-for-possible-climate-change-lies-by-new-york-attorney-general/. November 6, 2015.

Goklany, Indur M. 2009. "Deaths and Death Rates from Extreme Weather Events: 1900-2008." Jpands.org. http://www.jpands.org/vol14no4/goklany.pdf. Winter 2009.

Goldstein, Leo. 2016. "Who unleashed Climatism?" Wattsupwiththat. https://wattsupwiththat.com/2016/01/17/who-unleashed-climatism/. January 17, 2016.

—————. 2017. "President Trump: Time to Abolish Climate Alarmism in America." American Thinker. https://www.americanthinker.com/articles/2017/05/president_trump_time_to_abolish_climate_alarmism_in_america.html. June 1, 2017.

—————. 2018. "Summary of Science." Defyccc. https://defyccc.com/summary-of-science/. 2018.

Gonchar, Michael. 2018. "Young People Are Suing the Trump Administration over Climate Change. She's Their Lawyer." *New York Times.* https://www.

nytimes.com/2018/10/26/learning/learning-with-young-people-ar
e-suing-the-trump-administration-over-climate-change-shes-their-lawyer.
html. October 26, 2018.

Goodman, Ellen. 2007. "No Change in Political Climate." Archive.boston.
com. http://archive.boston.com/news/globe/editorial_opinion/oped/ar-
ticles/2007/02/09/no_change_in_political_climate/. February 9, 2007.

Gore, Albert. 1989. "An Ecological Kristallnacht. Listen." *New York Times*. https://
www.nytimes.com/1989/03/19/opinion/an-ecological-kristallnacht
-listen.html. March 19, 1989.

—————. 1992. *Earth in The Balance*. Rodale Books, Emmaus PA. http://
gelbspanfiles.com/wp-content/uploads/2013/06/ErintheB-360.jpg.
October 31, 2006.

—————. 2007. "Gore: Global warming fight needs China." China Daily.
February 7, 2007, 2007.

Grant, Natalie. 1998. "Green Cross: Gorbachev and Enviro-Communism."
The Register. http://americasurvival.org/wp-content/uploads/2017/06/
Global-Enviro-Communism.pdf. 1998.

Gray, William M. 1968. "Global View of the Origin of Tropical Disturbances
and Storms." Journals.ametsoc.org. https://journals.ametsoc.org/doi/
abs/10.1175/1520-0493%281968%29096%3C0669%3AGVOTOO%3
E2.0.CO%3B2. October 1, 1968.

Greenpeace. 2019. Greenpeace East Asia. Greenpeace.org. http://www.green-
peace.org/eastasia/. 2019.

Griswold, Alex. 2017. "Bill Nye Suggests Government Should Punish Parents
Who Have Too Many Children." Freebeacon. https://freebeacon.com/
culture/bill-nye-suggests-government-should-punish-parents-who
-have-too-many-children/. April 26, 2017.

GUFF. 2016. "15 Drastic Effects of Population Decline." Guff. https://guff.co
m/15-drastic-effects-of-population-decline. 2016.

GWPP. 2019. "Global Warming Petition Project." Petition Project. http://www.
oism.org/pproject/pproject.htm. 2019.

HadCM3 2018. "Hadley Centre Coupled Model, version 3." USGCRP.GCIS.
https://data.globalchange.gov/model/hadcm3. 2018.

Halperin, Ari. 2016. "Climate Alarmism and the Muzzling of Independent
Science" American Thinker. https://www.americanthinker.com/arti-
cles/2016/04/ climate_alarmism_and_the_muzzling_of_independent_
science.html. April 21, 2016.

—————. 2016. "The Paradoxical Origin of Climate Alarmism." American Thinker. https://www.americanthinker.com/articles/2016/02/the_paradoxical_origin_of_climate_alarmism.html. February 1, 2016.

Hampson, Spike. 2016. "Measuring Sea Level Is a Suspect Art." American Thinker. https://www.americanthinker.com/articles/2016/12/measuring_sea_level_is_a_suspect_art.html. December 5, 2016.

Hansen, James E. 2006. "Direct Testimony of James E. Hansen." Columbia.edu. http://www.columbia.edu/~jeh1/2007/IowaCoal_20071105.pdf. October 22, 2006.

—————. 2008. "Twenty years later: tipping points near on global warming." The Guardian. https://www.theguardian.com/environment/2008/jun/23/climatechange.carbonemissions. June 28, 2008.

Harper, Jim W. 2011. "Are Category 6 Hurricanes Coming Soon?" Scientific American. https://www.scientificamerican.com/article/are-category-6-hurricanes-coming/. August 23, 2011.

Harrington, Rebecca. 2015. "Here's how much of the world would need to be covered in solar panels to power Earth." Business Insider. https://www.businessinsider.com/map-shows-solar-panels-to-power-the-earth-2015-9. October 1, 2015.

Harris, Tom, and Timothy Ball. 2017. "Global Warming: Fake News from the Start." The Heartland Institute. https://www.heartland.org/news-opinion/news/global-warming-fake-news-from-the-start. December 20, 2017.

Harsanyi, David. 2017. "Bill Nye's View of Humanity Is Repulsive." The Federalist. https://thefederalist.com/2017/04/27/bill-nyes-view-of-humanity-is-repulsive/. April 27, 2017.

Harvey, Fiona. 2017a. "World has three years left to stop dangerous climate change, warns experts." Malabo Montpellier Panel. https://www.mamopanel.org/news/in-the-news/2017/jul/5/world-has-three-years-left-stop-dangerous-climate-/. July 5, 2017.

—————. 2017b. "World has three years left to stop dangerous climate change, warns experts." The Guardian. https://www.theguardian.com/environment/2017/jun/28/world-has-three-years-left-to-stop-dangerous-climate-change-warn-experts. June 28, 2017.

Hawkins, William R. 2010. "Global Warming and Cold War Thinking." American Thinker. https://www.americanthinker.com/articles/2010/03/global_warming_and_cold_war_th.html. March 14, 2010.

Heartland. 2018. "Climate Change Reconsidered" Non-Governmental Panel on Climate Change. http://climatechangereconsidered.org/wp-content/uplo

ads/2018/10/10-05-18-DRAFT-CCRII-Fossil-Fuels-Summary-for-Policy-Makers.pdf. 2018.

Helm, Dieter. 2017. "Cost of Energy Review." Biee.org. http://www.biee.org/wpcms/wp-content/uploads/Cost_of_Energy_Review.pdf. October 25, 2017.

Helmer, Roger. 2015. "The Big Green Lie - Fossil fuels are massively subsidized." Rogerhelmermep.wordpress.com. https://rogerhelmermep.wordpress.com/2015/05/27/the-big-green-lie-fossil-fuels-are-massively-subsidised/. May 27, 2015.

Hendricks, Bracken. 2010. "Don't believe in global warming? That's not very conservative." Washington Post. http://www.washingtonpost.com/wp-dyn/content/article/2010/11/05/AR2010110503155.html?hpid=opinionsbox1. November 7, 2010.

Hendrickson, Mark. 2018. "The End Game of Climate Change: Socialism." *Epoch Times.* https://www.theepochtimes.com/the-end-game-of-climate-change-socialism_2692791.html. October 30, 2018.

Henley, W. E. "Invictus – Poem by William Henley." poemhunter. https://www.poemhunter.com/poem/invictus/. 2019.

Herrera, Victor Manuel Velasco. 2008. "Scientist Warn that Earth will Enter 'Little Ice Age' Due to Decrease in Solar Activity." Banderas News. http://www.banderasnews.com/0808/eden-littleiceage.htm. August 2008.

Hickey, Colin, Travis N. Rieder, and Jake Earl 2016. "Population Engineering and the Fight against Climate Change." NPR.org. https://www.npr.org/documents/2016/jun/population_engineering.pdf. 2016.

Hinderaker, John. 2014. "The Epic Hypocrisy of Tom Steyer." Powerline. blog. https://www.powerlineblog.com/archives/2014/04/the-epic-hypocrisy-of-tom-steyer.php. April 20, 2014.

History. 1925. "Scopes Monkey trial." History. https://www.history.com/this-day-in-history/monkey-trial-begins. 1925.

HOC. 2010. "The disclosure of climate data from the Climatic Research Unit at the University of East Anglia." House of Commons Science and Technology Committee. https://publications.parliament.uk/pa/cm200910/cmselect/cmsctech/387/387i.pdf. March 24, 2010.

Hodge, Scott A. 2010. "Oil Industry Taxes: A Cash Cow for Government." TaxFoundation.org. https://taxfoundation.org/oil-industry-taxes-cash-cow-government/. July 28, 2010.

Hoven, Randell. 2012. "Global Warming Melts Away," American Thinker. https://www.americanthinker.com/articles/2012/05/global_warming_melts_away.html. May 3, 2012.

—————. 2018. "NASA's Rubber Ruler: An Update." American Thinker. https://www.americanthinker.com/articles/2018/01/nasas_rubber_ruler_an_update.html. January 1, 2018.

Howard, Jacqueline. 2017. "Is there a link between climate change and diabetes?" CNN. https://www.cnn.com/2017/03/20/health/climate-change-type-2-diabetes-study/. March 21, 2017.

Hughes, Austin L. 2012. "The Folly of Scientism." The New Atlantis. https://www.thenewatlantis.com/publications/the-folly-of-scientism. Fall 2012.

Hunter, Derek. 2017. "Unhinged: Liberals Suffing From Nightmares, Insomnia, Binge Eating Since Trump's Election." The Daily Caller. https://dailycaller.com/2017/03/18/unhinged-liberals-suffering-nightmares-insomnia-binge-eating-since-trumps-victory/. March 18, 2017.

Huston, Warner Todd. 2015. "Nailed It: ABC Predicted NYC Would be Under Water from Climate Change by 2015." Breitbart. https://www.breitbart.com/the-media/2015/06/13/nailed-it-abc-predicted-nyc-would-be-under-water-from-climate-change-by-2015/. June 13, 2015.

Hyde, Howard. 2015. "Climate Change: Where is the Science?" Rendevouswithdestiny.blogspot.com. http://rendevouswithdestiny.blogspot.com/ 2015/06/climate-change-where-is-science.html. June 11, 2015.

IBD. 2011. "Biden Endorses 'One Child Policy'." *Investor's Business Journal.* https://www.investors.com/politics/editorials/biden-endorses-one-child-policy/. August 23, 2011.

IBD. 2014. "A World In Chaos, And Kerry's Talking Climate Change." *Investor's Business Daily.* https://www.investors.com/in-a-world-on-fire-kerry-call s-climate-the-biggest-challenge/. August 14, 2014.

ICCER. 2010. "The Independent Climate Change E-mails Review." Cce.review. org. http://www.cce-review.org/pdf/FINAL%20REPORT.pdf. July 2010.

Id, Jeff. 2009. "Open Letter On Climate Legislation." Noconsensus. wordpress.com. https://noconsensus.wordpress.com/2009/11/13/open-letter/#comment-11917. November 13, 2009.

Indiana. 2018. "Milankovitch Cycles." Indiana.edu. http://www.indiana.edu/~geol105/images/ gaia_chapter_4/milankovitch.htm. 2018.

IPCC. 1990. "IPPC First Assessment Report." Intergovernmental Panel on Climate Change. https://www.ipcc.ch/report/ar1/wg1/. 1990.

—————. 1990. "The First Assessment Report (FAR)." International Panel on Climate Change. https://www.ipcc.ch/report/ar1/wg2/. 1990.

IPPC. 1995. "The Second Assessment Report (SAR)." International Panel on Climate Change. https://www.ipcc.ch/site/assets/uploads/2018/06/2nd-assessment-en.pdf. 1995.

IPPC. 2001. "The Third Assessment Report(s) (TAR)." International Panel on Climate Change. https://www.ipcc.ch/report/ar3/wg1/. 2001.

IPCC. 2009. "U.N. Conference on Climate Change." C-Span. https://www.c-span.org/video/?290744-1/un-conference-climate-change. December 16, 2009.

IPCC. 2013. United Nations Intergovernmental Panel on Climate Change. International Panel on Climate Change. http://www.climatechange2013.org/images/report/WG1AR5_ALL_FINAL.pdf. 2013.

IPCC. 2014. "AR5 Synopsis Report Climate Change Report 2014" IPCC.ch. https://www.ipcc.ch/report/ar5/syr/. 2014.

—————. 2014. "Fifth Assessment Report (AR5 2014)." International Panel on Climate Change. https://unfccc.int/topics/science/workstreams/cooperation-with-the-ipcc/the-fifth-assessment-report-of-the-ipcc. 2014.

IPCC. 2018. "Summary for Policymakers of IPCC Special Report on Global Warming of 1.5°C Approved by Government." IPCC. https://www.ipcc.ch/2018/10/08/summary-for-policymakers-of-ipcc-special-report-on-global-warming-of-1-5c-approved-by-governments/. October. 8, 2018.

IPCC-WGI. 2001. "The IPCC Working Group 1 (WPG)." Intergovernmental Panel on Climate Change. https://www.ipcc.ch/working-group/wg1/?idp=5. 2001.

IPPCSpecial. 2018. "Intergovernmental Panel on Climate Change." International Panel on Climate Change. https://report.ipcc.ch/sr15/pdf/sr15_spm_final.pdf. 2018.

Jayaraj, Vijay. 2017. "Blessing or Curse? The Curious Case of Carbon Dioxide." American Thinker. https://www.americanthinker.com/articles/2017/12/blessing_or_curse_the_curious_case_of_carbon_dioxide.html. December 22, 2017.

Jenkins, Holman W., Jr. 2014. "Jenkins: Personal Score-Settling Is the New Climate Agenda." *Wall Street Journal*. https://www.wsj.com/articles/holman-jenkins-personal-score-settling-is-the-new-climate-agenda-1393630082. February 28, 2014.

Jiechi, Yang. 2010. "Foreign Minister Yang Jiechi Answers Questions from Domestic and Overseas Journalists on China's Foreign Policy."

China-Embassy.org. http://sy.china-embassy.org/eng/xwfb/t662733.htm. March 8, 2010.

Johnson, Ian. 2017. "Climate change helped cause Brexit, says Al Gore." Independence. https://www.independent.co.uk/environment/brexit-climate-change-al-gore-says-global-warming-syria-war-h elped-leave-vote-a7645866.html. March 27, 2017.

Johnson, Paul. 2007. *Intellectuals: From Marx and Tolstoy to Sartre and Chomsky*. Harper Perennial, New York NY. https://www.amazon.com/Intellectual s-Marx-Tolstoy-Sartre-Chomsky/dp/0061253170. May 1, 2007.

Joondeph, Brian C. 2017a. "Is There Anything Climate Change Can't Do?" American Thinker. https://www.americanthinker.com/articles/2017/03/is_there_anything_climate_change_cant_do.html. March 29, 2017.

—————. 2017b. "When climate change warriors can't keep their stories straight." American Thinker. https://www.americanthinker.com/articles/2017/04/ when_climate_change_warriors_cant_keep_their_stories_straight_comments.html. April 3, 2017.

—————. 2017c. "More Fake News from the Climate Change Warriors." American Thinker. https://www.americanthinker.com/articles/2017/06/more_fake_news_from_the_climate_change_warriors.html. June 30, 2017.

—————. 2018d. "Follow the Money on Climate Caterwauling." American Thinker. https://www.americanthinker.com/articles/2018/10/follow_the_money_on_climate_caterwauling.html. October 24, 2018.

Jubilee. 2019. *Jubilee Debt Campaign*. JubileeDebt.org. https://jubileedebt.org.uk/. 2019.

Karl, Thomas R., Anthony Arguez, et al. 2015. "Possible Artifacts of Data Biases in the Recent Global Surface Warming Hiatus." *Science*. http://science.sciencemag.org/content/348/6242/1469. June 26, 2015.

Kaufman, Ari. 2007. "Global Whining." American Thinker. https://www.americanthinker.com/blog/2007/02/global_whining.html. February 16, 2007.

Kemp, Andrew C., Benjamin P. Horton, et al. 2010. "Climate related sea-level variations over the past two millennia." Wattsupwiththat.files.wordpress.com. https://wattsupwiththat.files.wordpress.com/ 2011/06/pnas_kemp-etal_2011_sea_level_rise.pdf. October 28, 2010.

Kennedy, Donald. 2007. "Year of the Reef." *Science*. http://science.sciencemag.org/content/318/5857/1695. December 14, 2007.

Kerr, Richard A. 1997. "Did a Blast of Sea-Floor Gas Usher in a New Age?" *Science*, 1267. http://science.sciencemag.org/content/275/5304/1267. February 28, 1997.

Kirby, Jasper, Joachim Curtius, et al. 2011. "Role of sulphuric acid, ammonia and galactic cosmic rays in atmospheric aerosol nucleation." Nature. https://www.nature.com/articles/nature10343. August 25, 2011.

Kirk, Russell. 2001. *The Conservative Mind*, Gateway Editions, Regnery Publishing Washington DC; 7th revised edition, chapter 11. https://www.amazon.com/Conservative-Mind-Burke-Eliot/dp/0895261715. September 1, 2001.

Kirkby, Jasper. 2009. "Cosmic rays and climate." CERN Document Server. http://cdsweb.cern.ch/record/1181073. June 4, 2009.

Klaus, Vaclav. 2008. "Climate alarmists pose real threat to freedom." Mannkal. http://mannkal.org/downloads/environment/climatealarmistspose.pdf. March 12, 2008.

Klein, Naomi. 2002. *No Logo*. Picador. https://www.goodreads.com/book/show/647.No_Logo. April 6, 2002.

Knorr, Bryce. 2019. "Corn Outlook—Keep hedging old crop corn." Farmprogress. https://www.farmprogress.com/story-weekly-corn-review-0-30766. March 25, 2019.

Ko, Lisa. 2016. "Unwanted Sterilization and Eugenics Programs in the United States." Pbs.org/independentlens. http://www.pbs.org/independentlens/blog/unwanted-sterilization-and-eugenics-programs-in-the-united-states/. January 29, 2016.

Kotecki, Peter. 2018. "10 countries at risk of becoming demographic time bombs." Business Insider. https://www.businessinsider.com/10-countries-at-risk-of-becoming-demographic-time-bombs-2018-8. August 8, 2018.

Kottasova, Ivana. 2015. "World poverty rate to fall below 10% for the first time." Money.cnn.com. https://money.cnn.com/2015/10/05/news/economy/poverty-world-bank/. October 5, 2015.

KPMG. 2017. "Agriculture in Africa." KPMGafrica. http://www.blog.kpmgafrica.com/agriculture-in-africa/. 2017.

Krisinger, Chris J. 2018. "Will Global Warming Destroy the World? Ask America's Farmers." American Thinker. https://www.americanthinker.com/articles/2018/10/ will_global_warming_destroy_the_world_ask_americas_farmers.html. October 26, 2018.

Kudla, John. 2018. "Things Your Professor Didn't Tell You About Climate Change." American Thinker. https://www.americanthinker.com/

articles/2018/02/ things_your_professor_didnt_tell_you_about_climate_change.html. February 3, 2018.

Kuenkel, Petra, et al. 2018. "The Club of Rome Climate Emergency Plan." Club of Rome. https://www.clubofrome.org/project/the-club-of-rome-climate-emergency-plan/. December 4, 2018.

Laframboise, Donna. 2010. "IPCC Says Climate Prediction Impossible." Nofrackingconsensus.com. https://nofrakkingconsensus.com/2010/06/30/ipcc-says-climate-prediction-impossible/. June 30, 2010.

LaFramboise, Donna. 2011. *The Delinquent Teenager who was Mistaken for the World's Top Climate Expert.* Ivy Press, Brighton UK, . https://www.amazon.com/Delinquent-Teenager-Mistaken-Climate-ebook/dp/B005UEVB8Q/ref=sr_1_1?ie=UTF8&qid=1374848997&sr=8-1&keywords=delinquent+teenager+ipcc. 2011.

Lamb, Christopher. 2014. "Pope Francis to publish encyclical on climate change." *International Catholic News Weekly.* https://www.thetablet.co.uk/news/1385/pope-francis-to-publish-encyclical-on-climate-change. November 13, 2014.

Laymann, Evan. 2015. "Obama Strikes First in War of Words with Congress over Global Warming." *Scientific American.* https://www.scientificamerican.com/article/obama-strikes-first-in-war-of-words-with-congress-over-global-warming/. January 21, 2015.

Lederman, Leon M. 2001. Biology Cabinet. Biocab. http://www.biocab.org/About_Us.html. 2001.

Leftcoast. 2019. Leftcoast Grassfed. LeftCoastGrassfed. https://leftcoastgrassfed.com/. 2019.

Leipzig. 2005. "The Leipzig Declaration on Global Climate Change." Henryhbauer.homestead.com. https://henryhbauer.homestead.com/Leipzig_DeclarationPontius2005.pdf. 2005.

Leuck, Dale. 2017. "Disingenuous Climate Science Debunked." American Thinker. https://www.americanthinker.com/articles/2017/03/disingenuous_climate_science_debunked.html. March 27, 2017.

—————. 2018. "Global Warming: The Evolution of a Hoax." American Thinker. https://www.americanthinker.com/articles/2018/03/global_warming_the_evolution_of_a_hoax.html. March 21, 2018.

Lewis, Avi, et al. 2014. "This Changes Everything – The Film." Thischangeseverything.com. https://thischangeseverything.org/the-documentary/. 2014.

Lewis, James. 2007a. "Is there an average global temperature?" American Thinker. https://www.americanthinker.com/articles/2007/03/is_there_an_average_global_tem_1.html. March 18, 2007.

—————. 2007b. "$ciense Mag Jumps on Global Moneywagon." American Thinker. https://www.americanthinker.com/articles/2007/12/cience_mag_jumps_on_global_mon.html. December 26, 2007.

Li, Peng, Changhui, Peng, et al. 2017. "Quantification of the response of global terrestrial net primary production to multifactor global change." Science Direct. https://www.sciencedirect.com/science/article/pii/S1470160X17300274. May 2017.

Lider, Julian. 1980. "The Correlation of World Forces: the Soviet Concept." Sage Journals. https://journals.sagepub.com/doi/abs/10.1177/002234338001700205. June 1, 1980.

Lilley, Peter. 2016. "£300 The Cost of the Climate Change Act." Thegwpf.org. https://www.thegwpf.org/content/uploads/2016/12/CCACost-Dec16.pdf. 2016.

Lindzen, Richard S. 1992a. "Global Warming The Origin and Nature of the Alleged Scientific Consensus." Cato Review of Business & Government. https://object.cato.org/sites/cato.org/files/serials/   files/regulation/1992/4/v15n2-9.pdf. Spring 1992.

—————. 1992b. "Global Warming: The Origin and Nature of the Alleged Scientific Consensus." Eaps.mit.edu. http://eaps.mit.edu/faculty/lindzen/153_Regulation.pdf. November 12, 2002.

—————. 2008a. "Biograph." Eaps.mit..edu. https://eapsweb.mit.edu/people/rlindzen. 2008.

—————. 2008b. "Climate Science: Is it currently designed to answer questions?" Cornell University. https://arxiv.org/abs/0809.3762. https://arxiv.org/abs/0809.3762. September 22, 2008.

—————. 2014. "Climate Science: Is it currently designed to answer questions?" Global Research. https://www.globalresearch.ca/climate-science-is-it-currently-designed-to-answer-questions/16330. September 22, 2014.

Liu, Na, Chen. Hong-Xia, Lu, Lian-Gang. 2005. "Teleconnection of IOD Signal in the Upper Troposphere over Southern High Latitudes." Journal of Oceanography, vol. 63 (2007), 155–5707. September 15, 2006.

Lofthus, Andre. 2014. "Global Warming and Settled Science." American Thinker. https://www.americanthinker.com/articles/2014/04/global_warming_and_settled_science.html. April 24, 2014.

Lomborg, Bjorn. 2001. "The Skeptical Environmentalist: Measuring the Real State of the World." Cambridge University Press. https://www.lomborg.com/skeptical-environmentalist. September 10, 2001.

—————. 2002. "Bjørn Lomborg's comments to the 11-page critique in January 2002 Scientific American (SA)." Static.scientificamerican.com. https://static.scientificamerican.com/ sciam/assets/media/pdf/lomborgrebuttal.pdf. February 31, 2002.

Long, Edward R. 2007. "The NOAA Database and Global Warming." American Thinker. https://www.americanthinker.com/articles/2017/10/the_noaa_database_and_global_warming.html. October 12, 2017.

—————. 2016. "An Inspection of NOAA's Global Historical Climatology NetWork Verson 3 Unadjusted Data." Solanqui. http://www.solanqui.com/An%20Inspection%20of%20NOAA%20V3%20Raw%20Data.pdf. 2016.

—————. 2017. "An Inspection of NOAA's Global Historical Climatology NetWork Verson 3 Unadjusted Data." https://www.heartland.org/publications-resources/publications/an-inspection-of-noaas-global-historical-climatology-network-verson-3-unadjusted-data. October 7, 2017.

Looker, Dan. 2016. "Corn, Soybean, Wheat, Price Projections for 2017-18 Marketing Year." Successful Farming. https://www.agriculture.com/markets/analysis/crops/corn-soybean-wheat-price-projections-for-2017-18-marketing-year. November 15, 2016.

Ludden, Jennifer. "Should We Be Having Kids in the Age of Climate Change?" NPR. https://www.npr.org/2016/08/18/479349760/should-we-be-having-kids-in-the-age-of-climate-change. August 18, 2016.

Ludwig, E. Jeffrey. 2017. "Trump Steps on the Paris Agreement, Stands for Sovereignty." American Thinker. https://www.americanthinker.com/articles/2017/06/ trump_steps_on_the_paris_agreement_stands_for_sovereignty.html. June 2, 2017.

—————. 2018a. "The UN Wants to be Our World Government By 2030." American Thinker. https://www.americanthinker.com/articles/2018/10/the_un_wants_to_be_our_world_government_by_2030.html. October 27, 2018.

—————. 2018b. "The 'Soft' but Real UN World Government Plan." American Thinker. https://www.americanthinker.com/articles/2018/11/the_soft_but_real_un_world_government_plan.html. November 21, 2018.

Malthus, Thomas. 1798. *An Essay on the Principle of Population*. Esp.org/books/ malthus. http://www.esp.org/books/malthus/population/malthus.pdf. 1798.

Mann, Michael E., Raymond S. Bradley, and Malcolm K. Hughes. 1998. "Global-scale temperature patterns and climate forcing over the past six centuries." Geo.umass.edu. http://www.geo.umass.edu/faculty/bradley/mann1998. pdf. 1998.

Masters, Jeff. 2019. "Hurricane Forecast Computer Models." Wunderground. https://www.wunderground.com/hurricane/models.asp. 2019.

McIntyre, Stephen. 2009. "Yamal: A 'Divergence' Problem." Climate Audit. https://climateaudit.org/2009/09/27/yamal-a-divergence-problem/. September 27, 2009.

McIntyre, Steven, and Ross McKitrick. 2003. "Corrections to the Mann et.al. (1998) Proxy Data Base and Northern Hemisphere Average Temperature Series." Multi-science.co.uk. http://www.multi-science.co.uk/ mcintyre-mckitrick.pdf. 2003.

McKitrick, Ross. 2010. "Understanding the Climategate Inquiries." Rossmckitrick.weebly.com. http://rossmckitrick.weebly.com/up-loads/4/8/0/8/4808045/rmck_climategate.pdf. September 2010.

McNicoll, Brian. 2018. "Time: Global Warming Must Be Stopped in 12 Years No Matter What." Accuracy in Media. https://www.aim.org/aim-column/ global-warming-must-be-stopped-in-12-years-no-matter-what/. October 9, 2018.

Meadows, Donella H., Dennis L. Meadows, et al. 1972. *The Limits to Growth*. Universe Books. https://www.clubofrome.org/report/the-limits-to-growth/. 1972.

Mendoza, Christopher. 2018. "The New Age of Coal." American Thinker. https://www.americanthinker.com/articles/2018/09/the_new_age_of_ coal.html. September 22, 2018.

Meotti, Giulio. 2018a. "Europe demographic suicide: See Greece." Arutz Sheva Israelnationalnews.com. http://www.israelnationalnews.com/Articles/ Article.aspx/22674. August 30, 2018.

—————2018b. "Europe is the world turned upside down." Arutz Sheva Israelnationalnews.com. http://www.israelnationalnews.com/Articles/ Article.aspx/22880. October 18, 2018.

Merrifield, D. Bruce. 2007. "Global Warming and Solar Radiation." American Thinker. https://www.americanthinker.com/articles/2007/07/global_ warming_and_solar_radia_1.html. July 11, 2007.

Mersarovic, Mihajlo, and Eduard Pestel. 1975. Mankind at the Turning Point. Hutchingson. https://www.clubofrome.org/report/mankind-at-the-turning-point/. 1975.

Michaels, Pat, and Chip Knappenberger. 2015. "Climate models versus climate reality." Judith Curry. https://judithcurry.com/2015/12/17/climate-models-versus-climate-reality/. December 17, 2015.

Mikkelson, David. 2008. "A Tale of Two Houses." Snopes. https://www.snopes.com/fact-check/tale-two-houses/. 2008.

Mikulska, Anne, and Eryk Kosinski. 2018. "Explaining Poland's Coal Paradox." *Forbes.* https://www.forbes.com/sites/thebakersinstitute/2018/03/28/explaining-polands-coal-paradox/#4b4b3afa4867. March 28, 2018.

Milankovitch, Milutin . 2019. Milankovitch Cycles. Indiana.edu. http://www.indiana.edu/~geol105/images/gaia_chapter_4/milankovitch.htm. 2019.

Miller, Brandon. 2017. "California's drought is almost over." CNN. https://www.cnn.com/2017/01/26/us/weather-california-drought/. January 26. 2017.

MIT. 2018. "Clausius-Clapeyron." http://web.mit.edu/16.unified/www/FALL/thermodynamics/notes/node64.html. 2018.

Monckton, Viscount. 2008. "Obama on the 'urgency' of combating 'global warming'." American Thinker. https://www.americanthinker.com/articles/2008/11/obama_on_the_urgency_of_combat.html. November 26, 2008.

Montford, Andrew. 2010. *The Climategate Inquiries.* Thegwpf.org. https://www.thegwpf.org/images/stories/gwpf-reports/Climategate-Inquiries.pdf. September 2010.

Mooney, Kevin. 2017. "CIA Veteran Sees Russian Connection to 2 Groups Opposing Fracking, Pipelines." *Daily Signal.* https://www.dailysignal.com/2017/08/25/cia-veteran-sees-russian-connection-to-2-groups-opposing-fracking-pipelines/. August 25, 2017.

Moore, Stephen, and Kathleen Hartnett. 2016. *Fueling Freedom: Exposing the Mad War on Energy.* Regnery Publishing, Washington DC. https://www.amazon.com/Fueling-Freedom-Exposing-Mad-Energy/dp/1621574091. May 23, 2016.

Moran, Rick. 2007. "Bali Climate Conference Ignores Dissenter." American Thinker. https://www.americanthinker.com/blog/2007/12/bali_climate_conference_ignore.html. December 9, 2007.

————. 2008. "Rank hypocrisy from George Soros." American Thinker. https://www.americanthinker.com/blog/2008/08/rank_hypocrisy_from_george_sor.html. August 21, 2008.

Morano, Marc. 2010. "More Than 1000 International Scientists Dissent Over Man-Made Global Warming Claims—Challenge UN IPCC & Gore." https://www.climatedepot.com/2010/12/08/special-report-more-than-1000-international-scientists-dissent-over-manmade-global-warming-claims-challenge-un-ipcc-gore-2/. December 8, 2010.

Morrison, Spencer P. 2017. "Wind and Solar Energy Are Dead Ends." American Thinker. https://www.americanthinker.com/articles/2017/07/wind_and_solar_energy_are_dead_ends.html. July 12, 2017.

Nahle, Nasif. 2007. "Geologic Global Climate Change." Biocab.org. http://www.biocab.org/Carbon_Dioxide_Geological_Timescale.html. March 14, 2007.

NAS. 2012. "A National Strategy for Advancing Climate Modeling." Nas.edu. http://dels.nas.edu/Report/National-Strategy-Advancing-Climate/13430. 2012.

NASA. 2004. "An introduction to general relativity." NASA press kit. https://spaceflightnow.com/delta/d304/040417theory.html. April 12, 2004.

—————. 2005. "NASA—What's the Difference Between Weather and Climate?" https://www.nasa.gov/mission_pages/noaa-n/climate/climate_weather.html. February 1, 2005.

—————. 2014. "Solar Mini-Max." NASA Science Share the Science. https://science.nasa.gov/science-news/science-at-nasa/2014/10jun_solarminimax/. June 10, 2014.

—————. 2015a *"NASA, NOAA Find 2014 Warmest Year in Modern Record."* https://www.nasa.gov/press/2015/january/nasa-determines-2014-warmest-year-in-modern-record. January 16, 2015.

—————. 2015b. "NASA Study: Mass Gains of Antarctic Ice Sheet Greater than Losses." Nasa.gov. https://www.nasa.gov/feature/goddard/nasa-study-mass-gains-of-antarctic-ice-sheet-greater-than-losses. October 30, 2015.

—————. "Dark Energy, Dark Matter." NASA Science Share the Science. https://science.nasa.gov/astrophysics/focus-areas/what-is-dark-energy/. 2018.

—————. 2019. "Climate Models." https://www.climate.gov/maps-data/primer/climate-models. 2019.

Nastu, Paul. 2008. "James Hansen: Try Fossil Fuel CEOs For 'High Crimes Against Humanity'." Environmental Leaders. https://www.environmentalleader.com/2008/06/james-hansen-try-fossil-fuel-ceos-for-high-crimes-against-humanity/. June 24, 2008.

National Academy. 2019. "About the Climate Communications Initiative." National Academy. http://sites.nationalacademies.org/sites/climate/SITES_191032. 2019.

NEE. 2017a. "A 100% Solar-Powered Future Is Impossible—Requires 7.2 Times More Silver Than Currently Exists." National Economics Editorial. https://nationaleconomicseditorial.com/2017/06/05/solar-powered-future-impossible/. June 5, 2017.

—————. 2017b. "Wind Energy Meets Just 0.46% Of Global Energy Demand—Despite Hundreds Of Billions In Investment." National Economics Editorial. https://nationaleconomicseditorial.com/2017/06/26/wind-power-future-impossible/. June 26, 2017.

Neslen, Arthur. 2016. "Renewable energy smashes global records in 2015, report shows." TheGuardian.com. https://www.theguardian.com/environment/2016/jun/01/renewable-energy-smashes-global-records-in-2015-report-shows. June 1, 2016.

Neudorff, Brian. 2007. "Climate Controversy: TV Meteorologist vs. Dr. Heidi Cullen, Weather Channel Climate Expert." WX-MAN's Musings. http://wx-man.com/blog/?p=500. January 22, 2007.

Nextgrandminimum.com. https://nextgrandminimum.com/2014/12/15/mini-ice-age-2015-2035-top-scientists-predict-global-cooling-2015-2050/. December 15, 2014.

Nickel, Rob. 2017. "Special Report: Drowning in grain - How Big Ag sowed seeds of a profit-slashing glut." Reuters.com. https://www.reuters.com/article/us-grains-supply-special-report/special-report-drowning-in-grain-how-big-ag-sowed-seeds-of-a-profit-slashing-glut-idUSKCN1C21AR. September 27, 2017.

Nielsen, Aly, and Joseph Rossell. 2015. "Katrina Anniversary: Media's 10 most Outlandish Predictions Full of Hot Air." MRS NewsBuster. https://www.newsbusters.org/blogs/business/aly-nielsen/2015/08/26/katrina-anniversary-medias-10-most-outlandish-hurricane. August 26, 2015.

NIPCC. 2009. Climate Change Reconsidered (2009). http://climatechangereconsidered.org/climate-change-reconsidered-2009-nipcc-report/. June 2009.

—————. 2014. Climate Change Reconsidered II (2013). http://climatechangereconsidered.org/climate-change-reconsidered-ii-biological-impacts/. April 9, 2014.

—————. 2018. "Climate Change Reconsidered II: Fossil Fuels." Non-Governmental Panel on Climate Change. http://climatechangereconsidered.org/climate-change-reconsidered-ii-fossil-fuels/. December 4, 2018.

NOAA. 2019. NOAA's the Regional Climate Centers. Climate Prediction Center. https://www.cpc.ncep.noaa.gov/products/analysis_monitoring/regional_monitoring/usa.shtml, 2019.

Nordhaus, William D. 2012. "Why the Global Warming Skeptics are Wrong." *New York Review of Books*. https://www.nybooks.com/articles/2012/03/22/why-global-warming-skeptics-are-wrong/. February 22, 2012.

NRC. 2012. "Research Universities and the Future of America." Nap.edu. https://www.nap.edu/catalog/13396/research-universities-and-the-future-of-america-ten-breakthrough-actions. 2012.

Nye, Bill. 2017. Bill Nye Suggests POPULATION CONTROL to stop Climate Change—Bill Nye Saves the World. YouTube.com. https://www.youtube.com/watch?v=SZfT5MgSbDQ. April 26, 2017.

NYT 2014. "Meager Returns for the Democrats Biggest Donor -Tom Steyer." *New York Times*. https://www.nytimes.com/2014/11/07/us/politics/-meager-returns-for-the-democrats-biggest-donor-tom-steyer.html. November 6, 2014.

Obama, Barack. 2008. "A New Chapter on Climate Change." YouTube.com. https://www.youtube.com/watch?v=h-vG2XptIEJk&eurl=http%3A%2F%2Fvoices.washingtonpost.com%2Fthe-trail%2F2008%2F11%2F18%2Fobama_sends_a_message_to_gover.html. November 17, 2008.

—————. 2015. 2015 State of the Union address. ABCNews.go.com. https://abcnews.go.com/Politics/state-union-2015-fact-check-obamas-rhetoric-reality/story?id=28355350. January 20, 2015.

OCT. 2019. Our Children's Trust. https://www.ourchildrenstrust.org/mission-statement/. 2019.

OECD. 2015. "The Africa Competitiveness Report 2015." Weforum.org. http://www3.weforum.org/docs/WEF_ACR_2015/Africa_Competitiveness_Report_2015.pdf. 2015.

Olsen, Roger Fjellstad. 2018. "What will happen if the carbon dioxide content of atmosphere is as high as 300 parts per million?" Quora. https://www.quora.com/What-will-happen-if-the-carbon-dioxide-content-of-atmosphere-is-as-high-as-300-parts-per-million. February 3, 2017.

Oregon. 2007. "Oregon Petition." Oism.org. https://defyccc.com/oregon-petition/. 2017.

Oreskes, Naomi. 2006. "Charney and MacDonald (JASON) committees." Stanford.edu. https://web.stanford.edu/dept/cisst/ORESKES.Senate%20 EPW.FINAL.pdf.

Orwell, George. 1945. "Animal Farm: A Fairy Story." Orwell.ru. http://www. orwell.ru/library/novels/Animal_Farm/english/eaf_go. 1945.

—————. 1946. "Politics and the English Language." George Orwell: Politics. http://www.orwell.ru/library/essays/politics/english/e_polit. 1946.

Osborn, Fairfield. 1948. *Our Plundered Planet*. Little, Brown, 1st edition. https://www.amazon.com/Our-Plundered-Planet-Fairfield-Osborn/ dp/0316666084. January 1948.

Osorio, Ivan. 2007. "O'Leary casts an bleary eye on 'Eco-nutters'." Competitive Enterprise Institute. https://cei.org/blog/oleary-casts-bleary-eye-eco-nutters. September 17, 2007.

Owens, Eric. 2014. "US College Professor Demands Imprisonment for Climate-Change Deniers." The Daily Caller. https://dailycaller.com/2014/03/17/u-s-college-professor-demands-imprisonment-for-climate-change-deniers/. March 17, 2014.

Palomar. 2013. Climate Change and Human Evolution 2013. Palomar.edu. https://www2.palomar.edu/anthro/homo/homo_3.htm. 2013.

Palomar. 2018. "Climate Change and Human Evolution." Palomar.edu. https:// www2.palomar.edu/anthro/homo/homo_3.htm. 2018.

Park, Madison. 2015. "Obama: No greater threat to future than climate change." CNN. https://www.cnn.com/2015/01/21/us/climate-change-us-obama/. January 21, 2015.

Parkenson, Giles. 2014. "Graph of the Day: Why 'experts' get it wrong on wind and solar." Renew Economy—Fair Dinkum Power News & Analysis. https://reneweconomy.com.au/graph-of-the-day-why-experts-get-it-wr ong-on-wind-and-solar-58816/. August 11, 2014.

Patterson, Matt. 2016. "What Lies Beneath." American Thinker. https://www. americanthinker.com/articles/2016/05/what_lies_beneath.html. May 17, 2016.

Patton, Vince. 2007. "Oregon Gov. to Fire State Climatologist for Saying Global Warming Not Due to Human Activity." Scot.net. https://www.sott.net/ article/126862-Oregon-Gov-to-Fire-State-Climatologist-for-Saying-Glo bal-Warming-Not-Due-to-Human-Activity. February 7, 2007.

Peracchio, Carol. 2011. "Pavlov's Voters." American Thinker. https://www.amer-icanthinker.com/articles/2011/06/pavlovs_voters.html. June 27, 2011.

Perry, Mark J. 2015. "18 spectacularly wrong apocalyptic predictions." Aei. org. https://www.aei.org/publication/18-spectacularly-wrong-apocal yptic-predictions-made-around-the-time-of-the-first-earth-day-in-19 70-expect-more-this-year-2/. April 21, 2015.

Pielke, Rogers, Jr. 2010. "Frank Press, President of the NAS, 1989 on Climate Policy." Roger Pielke, Jr. blog. http://rogerpielkejr.blogspot.com/2010/08/ frank-press-president-of-nas-1989-on.html. August 5, 2010.

Planck. 1900. "Planck's Law." Scienceworld. Wolfram.com, http://scienceworld. wolfram.com/physics/PlanckLaw.html. 1900.

Plautz, Jason. 2014. "The Climate Change Solution No One Will Talk About." *The Atlantic*. https://www.theatlantic.com/health/archive/2014/11/th e-climate-change-solution-no-one-will-talk-about/382197/. November 1, 2014.

Plimer, Ian. 2009. *Heaven and Earth: Global Warming: The Missing Science*. Quartet Books, London UK. https://www.goodreads.com/book/ show/6495152-heaven-and-earth. May 2009.

Poon, Linda. 2015. "There Are 200 Million Fewer Hungry People Than 25 Years Ago." NPR. https://www.npr.org/sections/goatsand- soda/2015/06/01/411265021/there-are-200-million-fewe r-hungry-people-than-25-years-ago. June 1, 2015.

Power, Samantha. 2003. "How to Kill a Country." The Atlantic. https://www.the- atlantic.com/magazine/archive/2003/12/how-to-kill-a-country/302845/. December 2003.

Prois, Jessica. 2015. "Voluntary Birth Control Is a Climate Change Solution Nobody Wants to Talk About." Huffington Post. https://www.huffingtonpost. ca/entry/birth-control-climate-change_us_565339cde4b0258edb322194. December 10, 2015.

Ranker. 2018. "Celebrities Who Live(d) in Malibu." Ranker. https://www. ranker.com/list/celebrities-who-live-in-malibu/celebrity-lists. 2018.

Reuters. 1981. "Barbara Ward, British Economist, Dies." *New York Times*. https://www.nytimes.com/1981/06/01/obituaries/barbara-ward-britis h-economist-dies.html. June 1, 1981.

—————. 2010. "Coverage of climate summit called short on science." *Washington Post*. http://www.washingtonpost.com/wp-dyn/content/arti- cle/2010/11/14/AR2010111404444.html?noredirect=on. November 15, 2010.

Revkin, Andrew C. 2012. "Peter Gleick Admits to Deception in Obtaining Heartland Climate Files." Dotearth.blog.nytimes.com. https://dotearth.

blogs.nytimes.com/2012/02/20/peter-gleick-admits-to-deception-in-obtaining-heartland-climate-files/. February 20, 2012.

Richman, Howard, and Raymond Richman. 2013. "Ten Year Anniversary of the Climate Change Paradigm Shift." American Thinker. https://www.americanthinker.com/articles/2013/08/ ten_year_anniversary_of_the_climate_change_paradigm_shift.html. August 21, 2013.

Ridley, Matt. 2017. "Wind Energy Meets Just 0.46% Of Global Energy Demand—Despite Hundreds Of Billions In Investment." The Spectator. https://www.spectator.co.uk/2017/05/wind-turbines-are-neither-clean-nor-green-and-they-provide-zero-global-energy/. May 13, 2017.

Rinehart, Rick. 2012. "Scientists Behaving Badly." American Thinker. https://www.americanthinker.com/articles/2012/02/scientists_behaving_badly_comments.html. February 25, 2012.

Rinkesh, Kukreja. 2018. "Serious Effect of Global Warming." Conserver Energy Future. https://www.conserve-energy-future.com/globalwarmingeffects.php. 2018.

Roberts, David. 2006. "An excerpt from a new book by George Monbiot." Grist.org. https://grist.org/article/the-denial-industry/. September 20, 2006.

Robinson, Arthur B., Noah E. Robinson, and Willie Soon. 2007. "Environmental Effects of Increased Atmospheric Carbon Dioxide." Journal of American Physicians and Surgeons. 2007.

Rogers, Norman. 2011. "Adventures in the Climate Trade." American Thinker. https://www.americanthinker.com/articles/2011/05/adventures_in_the_climate_trad.html. May 1, 2011.

—————. 2013a. "Science in the Service of Ideology: The National Climate Assessment." American Thinker. https://www.americanthinker.com/articles/2013/05/ science_in_the_service_of_ideology_the_national_climate_assessment.html. May 31, 2013.

—————. 2013b. "American Geophysical Union Scraps Science, Now Faith Based." American Thinker. https://www.americanthinker.com/articles/2013/06/american_geophysical_union_scraps_science_now_faith_based.html. June 29, 2013.

—————. 2013c. "Global Warming as Faith." American Thinker. https://www.americanthinker.com/articles/2013/08/global_warming_as_faith.html. August 6, 2013.

—————. 2013d. "The Climate-Industrial Complex." American Thinker. https://www.americanthinker.com/articles/2013/09/the_climate-industrial_complex.html. September 27, 2013.

—————. 2014. "The Corruption of Science." American Thinker. https://www.americanthinker.com/articles/2014/05/the_corruption_of_science.html. May 24, 2014.

—————. 2015a. "Pareto Speaks to Us About Environmentalism." American Thinker. https://www.americanthinker.com/articles/2015/03/pareto_speaks_to_us_about_environmentalism.html. March 27, 2015.

—————. 2015b. "Global Warming: Making the Ruling Class into the Crackpot Class." American Thinker. https://www.americanthinker.com/articles/2015/10/ global_warming_making_the_ruling_class_into_the_crackpot_class.html. October 10, 2015.

—————. 2017a. "The Global Warming Smoking Gun." American Thinker. https://www.americanthinker.com/articles/2017/01/the_global_warming_smoking_gun.html. January 16, 2017.

—————. 2017b. "Tell a Big Lie and Keep Repeating It." American Thinker. https://www.americanthinker.com/articles/2017/11/tell_a_big_lie_and_keep_repeating_it.html. November 11, 2017.1

—————. 2018a. "Parade of Impending Catastrophes." American Thinker. https://www.americanthinker.com/articles/2018/04/the_parade_of_impending_catastrophes.html. April 11, 2018.

—————.. 2018b. *Dumb Energy – A Critique of Wind and Solar Energy.* Dumb Energy Publishing; 2nd edition. https://www.amazon.com/Dumb-Energy-Critique-Solar-energy/dp/1732537631. July 23, 2018.

—————. 2018c. "Wind and Solar Energy: Good for Nothing." American Thinker. https://www.americanthinker.com/articles/2018/08/wind_and_solar_energy_good_for_nothing.html. August 11, 2018.

—————. 2018d. "Dumb Energy Advances in Colorado." American Thinker. https://www.americanthinker.com/articles/2018/09/dumb_energy_advances_in_colorado.html. September 24, 2018.

—————. 2018e. "Green Energy is the Perfect Scam Rogers." American Thinker. https://www.americanthinker.com/articles/2018/11/green_energy_is_the_perfect_scam.html. November 12, 2018.

Romm, Joe. 2014. "John Kerry Calls Climate Change 'World's Most Fearsome' Weapon Of Mass Destruction." ThinkProgress.com. https://thinkprogress.org/john-kerry-calls-climate-change-worlds-most-fearsome-weapon-of-mass-destruction-35403a4ceef8/. February 17, 2014.

Rose, David. 2013. "Global warming stopped 16 years ago, Met Office report reveals: MoS got it right about warming … so who are the 'deniers' now?" *Daily Mail.* https://www.dailymail.co.uk/news/article-2261577/

Global-warming-stopped-16-years-ago-Met-Office-report-reveal
s-MoS-got-right-warming--deniers-now.html?ito=feeds-newsxml. January
12, 2013.

—————. 2015. "NASA Climate Scientists: We said 2014 was the warmest
year on record ... but we're only 38%sure we were right." *Daily Mail.* https://
www.dailymail.co.uk/news/article-2915061/Nasa-climate-scientists-s
aid-2014-warmest-year-record-38-sure-right.html. January 17, 2015.

—————. 2017. "Exposed: How world leaders were duped into investing billions over manipulated global warming data." *Daily Mail.* https://
www.dailymail.co.uk/sciencetech/article-4192182/World-leaders-dupe
d-manipulated-global-warming-data.html. February 4, 2017.

Russell, Cook. 2014. "Smearing Climate Skeptics." American Thinker. https://
www.americanthinker.com/articles/2014/03/smearing_climate_skeptics.
html. March 27, 2014.

SA. 2006. "Mysterious Stabilization of Atmospheric Methane May Buy Time in
Race to Stop Global Warming," *Geophysical Research Letters*, November 23,
2006. Mysterious Stabilization of Methane, *Scientific American*, November
21, 2006.

Sadar, Anthony J. 2011. "Has Progressivism Ruined Environmental Science?"
American Thinker. https://www.americanthinker.com/articles/2011/08/
has_progressivism_ruined_environmental_science.html. August 22,
2011.

—————. 2012. "The Perspective of a Lifetime on Atmospheric Modeling."
American Thinker. https://www.americanthinker.com/articles/2012/08/
the_perspective_of_a_lifetime_on_atmospheric_modeling.html. August
4, 2012.

—————. 2014. "Reflections on the State of Climate Science." American
Thinker. https://www.americanthinker.com/articles/2014/07/reflections_
on_the_state_of_climate_science.html. July 24, 2014.

—————. 2016. "The Model Atmosphere and Global Warming," American
Thinker. https://www.americanthinker.com/articles/2016/04/the_
model_atmosphere_and_global_warming.html. April 12, 2016.

Salvaterra, Neanda. 2018. "Coal Shows Resilience in Global Comeback."
Wall Street Journal. https://www.wsj.com/articles/why-coals-powe
r-persists-1535976000. September 3, 2018.

Santayana, George. 2013. "1970's Global Cooling Alarmism." Popular Technology.
http://www.populartechnology.net/2013/02/the-1970s-global-cooling-
alarmism.html. February 28, 2013.

Santer, Benjamin D. 2018. Benjamin D. Santer. https://en.wikipedia.org/wiki/Benjamin_D._Santer. 2018.

Schimdt, Gavin. 2014. "Can we make better graphs of global temperature history." Realclimate.org. http://www.realclimate.org/index.php/archives/2014/03/can-we-make-better-graphs-of-global-temperature-history/. March 13, 2014.

Schlosberg, Mark. 2017. "Off Act is a Climate Changer." EcoWatch.com. https://www.ecowatch.com/off-act-tulsi-gabbard-2479880368.html. September 7, 2017.

Schmitt, Jerome J. 2007a. "Numerical Models, Integrated Circuits, and Global Warming." American Thinker. https://www.americanthinker.com/articles/2007/02/numerical_models_integrated_ci.html. February 28, 2007.

—————. 2007b. "Galileo Denied Consensus." American Thinker. https://www.americanthinker.com/articles/2007/04/galileo_denied_consensus.html. April 18, 2007.

—————. 2008a. "Liberals and Mathematical Models." American Thinker. https://www.americanthinker.com/articles/2008/01/liberals_and_mathematical_mode_1.html. January 9, 2008.

—————. 2008b. "Climate totalitarians." American Thinker. https://www.americanthinker.com/blog/2008/03/climate_totalitarians.html. March 12, 2008.

—————. 2008c. "'Grantsmanship' Distorts Global Warming Science." American Thinker. https://www.americanthinker.com/articles/2008/05/grantsmanshipand_the_global_wa.html. May 21, 2008.

—————. 2008d. "Corrupted science revealed." American Thinker. https://www.americanthinker.com/blog/2008/09/corrupted_science_revealed.html. September 24, 2008.

Schneiderman, A. G., et al. 2016. "Press Conference." Ag.ny.gov/press-release. https://ag.ny.gov/press-release/ag-schneiderman-former-vice-president-al-gore-and-coalition-attorneys-general-across. March 29, 2016.

Science. 1997 *Science*, Feb 28 (1997), 1267; *Geothermal Geophysical, Geosystems* 10 (2001), 1029.

Science. 2005 (August 26, 2005); *Science* (February 28, 1997), 1267; *Science News* (February 3, 2007), vol. 171, 67; *Science*, vol. 312 (June 9, 2006), 1485–89. http://www.Skymetrics.US/background/glossey.php -21k

Science. 2006. (June 6, 2006), 1454; A. V. Federov, P. S. Dekens, et al, The Pliocene Paradox, *Science*, vol. 312 (June 9, 2006), 1485–89; D. R. MacAyeal, Dept. Geophysical Sciences, Univ. of Chicago, http://www.

Geosci.uchicago.edu, National Academic Press. https://www.nap.edu/read/9755/chapter/1. January 2000. Science et al. 2000. "Reconciling observations of global temperature change."

SCOTUS. 2014. "SCOTUS—Obergefell vs. Hodges." Justice.gov. https://www.justice.gov/sites/default/files/crt/legacy/2015/06/26/obergefellhodgesopinion.pdf. 2014.

—————. 2015. "*Juliana v. United States Youth Climate Lawsuit.*" Supreme Court of the United States. https://www.supremecourt.gov/DocketPDF/18/18-505/67251/20181017183026537_In 20re%20United%20States%20%20-%20Petition%20for%20Mandamus.pdf. 2015.

—————. 2018. "Supreme Court Stays Juliana Suit." Pace Law Library. https://lawlibrary.blogs.pace.edu/2018/10/22/supreme-court-stays-juliana-suit/. October 22, 2018.

—————. 2018. "UNITED STATES et al. V. USDC OR." Supremecourt.gov. https://www.supremecourt.gov/orders/courtorders/073018zr_8mjp.pdf. 2018.

Seager, John. 2018. "Lower Birth Rate Trigger False Alarm." Populationconnection.com. https://www.populationconnection.org/false-alarms/. September 4, 2018.

Segalstad, Tom V. 2018. "Web info about CO_2 and the asserted 'Greenhouse Effect' Doom." Co2web.info. http://www.co2web.info/. 2018.

Seitz, Frederick. 1996. "A Major Deception on 'global warming'." *Wall Street Journal*. http://stephenschneider.stanford.edu/Publications/PDF_Papers/WSJ_June12.pdf. June 12, 1996.

SEM. 2006. "Surface Temperature Reconstructions fo the Last 2,000 Years." National Academy of Science Engineering Medicine. https://www.nap.edu/read/11676/chapter/3. 2006.

Sessions, Jeff. 2018. "Nationwide Injunctions Are a Threat to Our Constitutional Order." *National Review*. https://www.nationalreview.com/2018/03/nationwide-injunctions-stop-elected-branches-enforcing-law/. March 10, 2018.

Severinghouse. 1999, *Science* 286 (October 29, 1999), 930–34. October 29, 1999.

Seymour, Julia A. *2015*. "And That's the Way It Was: In 1972, Cronkite Warned of 'New Ice Age'" *MRC NewsBuster*. https://www.newsbusters.org/blogs/julia-seymour/2015/03/05/and-thats-way-it-was-1972-cronkite-warned-new-ice-age. March 5, 2015.

Shaviv, Nir J. 2006. "Cosmic Rays and Climate." Science Bits. http://www.sciencebits.com/CosmicRaysClimate.

—————. 2006. "The Milky Way Galaxy's Spiral Arm and Ice Age Epochs and the Cosmic Ray Connection." ScienceBits.com. 2006.

Sheahen, Thomas P. 2016. "The Peer Review Problem." American Thinker. https://www.americanthinker.com/articles/2016/11/the_peer_review_problem.html. November 28, 2016.

Sheen, Fulton. 1979. The Life of All Living. Galilee Trade, 1st edition. https://www.amazon.com/Life-All-Living-Fulton-Sheen/dp/0385154585/ref=sr_1_1?s=books&ie=UTF8&qid=1510520681&sr=1-1&keywords=The+life+of+all+living. August 7, 1979.

Sheppard, Marc. 2007. "Global Warmists Exploit Holovaust." American Thinker. https://www.americanthinker.com/articles/2007/11/green_fever_global_warming_and.html. November 30, 2007.

—————. 2009a. "The Evidence of Criminal Fraud." American Thinker. https://www.americanthinker.com/articles/2009/11/the_evidence_of_climate_fraud.html. November 21, 2009.

—————. 2009b. "Understanding Climategate's Hidden Decline." American Thinker. https://www.americanthinker.com/articles/2009/12/understanding_climategates_hid.html. December 6, 2009.

—————. 2010a. "Climategate: CRU Was But the Tip of the Iceberg."American Thinker. https://www.americanthinker.com/articles/2010/01/climategate_cru_was_but_the_ti.html. January 22, 2010.

—————. 2010b. "IPCC: International Pack of Climate Crooks." American Thinker. https://www.americanthinker.com/articles/2010/02/ipcc_international_pack_of_cli.html.

—————. 2010c. "Climategate's Phil Jones Confesses to Climate Fraud." American Thinker. https://www.americanthinker.com/articles/2010/02/climategates_phil_jones_confes.html. February 14, 2010.

—————. 2010d. "Climategate: One Year and Sixty House Seats Later," American Thinker. https://www.americanthinker.com/articles/2010/11/climategate_one_year_and_sixty_house_seats_later.html. November 17, 2010.

Sheppard, Noel. 2007. "Weapons of Global Warming Destruction." American Thinker. https://www.americanthinker.com/articles/2007/02/weapons_of_global_warming_dest.html. February 12, 2007.

Siber, John R. 1973. "The Humanities, Viewed as a Crucible for Higher Educations." New York Times. https://www.nytimes.com/1973/01/08/archives/the-humanities-viewed-as-a-crucible-for-higher-education-alienation.html. January 8, 1973.

Siegel, Josh. 2018. "Supreme Court pauses kids' climate change suit before trial." Washington Examiner. https://www.washingtonexaminer.com/policy/energy/supreme-court-pauses-kids-climate-change-suit-before-trial. October 19, 2018.

Singer, S. Fred. 2012a. "Climate Deniers Are Giving Us Skeptics a Bad Name." American Thinker. https://www.americanthinker.com/articles/2012/02/climate_deniers_are_giving_us_skeptics_a_bad_name.html. February 29, 2012.

—————. 2012b. "Climate Realism." American Thinker. https://www.americanthinker.com/articles/2012/09/climate_realism.html. September 26, 2012.

—————. 2014. "Cause of Pause in Global Warming." American Thinker. https://www.americanthinker.com/articles/2014/12/cause_of_pause_in_global_warming.html. December 29, 2014.

Sobel, Adam. 2015. "Did climate change cause California drought?" CNN. https://www.cnn.com/2015/04/08/opinions/sobel-california-drought/. April 8, 2015.

Sobieski, Daniel John. 2017. "Bill Nye, the Eugenics Guy." American Thinker. https://www.americanthinker.com/articles/2017/05/bill_nye_the_eugenics_guy.html. May 3, 2017.

Soon, Willie, and Sallie Baliunas. "Global Warming." Sage Journals. https://journals.sagepub.com/doi/abs/10.1191/0309133303pp391pr. September 1, 2003.

Spencer, Roy. 2008. "The Pacific Decadal Oscillation (PDO): Key to the Global Warming Debate?" Drroyspencer.com. http://www.drroyspencer.com/global-warming-background-articles/the-pacific-decadal-oscillation/. 2008.

—————. 2014. "ISH Surface Weather Stations Reporting 4+ times per day, 1986 thru 2009." Drroyspencer.com. http://www.drroyspencer.com/wp-content/uploads/ISH-station-map-1986-thru-2009-6-hrly.jpg. 2014.

—————. 2018a. Roy Spencer. Droyspencer.com. http://www.drroyspencer.com/global-warming-natural-or-manmade/. 2018.

—————. 2018b "The Pacific Decadal Oscillation (PDO) Key to the Global Warming Debate?" DRRoySpencer.com. http://www.drroyspencer.com/global-warming-background-articles/the-pacific-decadal-oscillation/. 2018.

Spry, Jamie. 2018. "PEER REVIEW SCIENCE: The Medieval Warming Period Was Indeed Global And Warmer Than Today." Climatism, https://climatism.blog/category/medieval-warm-period/. December 3, 2018.

St. John, Paige. 2017. "Skiers hit the slopes in bikini tops as California's endless winter endures a heat wave." Los Angeles Times. https://www.latimes.com/local/lanow/la-me-endless-winter-20170617-story.html. June 20, 2017.

Staff. 2007. "Macro vs. Micro Evolution." The Forerunner. http://forerunner.com/forerunner/X0737_Macro_vs._Micro_Evol.html. December 22, 2007.

Stanford. 2018. "Philosophy of Science." Stanford.edu. https://web.stanford.edu/class/symsys130/Philosophy%20of%20science.pdf. 2018.

Steele, Russ. 2014. "Mini Ice Age 2015-2035 ! Top Scientists Predict Global Cooling 2015-2050." NextGrandminimum.com. https://nextgrandminimum.com/2014/12/15/mini-ice-age-2015-2035-top-scientists-predict-global-cooling-2015-2050/. December 15, 2014.

Street, Chris. 2015. "Hello Al Gore: Low Sun Spot Cycle Could Mean another 'Little Ice Age'." American Thinker. https://www.americanthinker.com/articles/2015/05/ hello_al_gore_low_sun_spot_cycle_could_mean_another_little_ice_age.html. May 6, 2015.

Strong, Maurice. 2000. *Where on Earth are We Going?* Good Reads, San Francisco CA. https://www.goodreads.com/book/show/1257555.Where_On_Earth_Are_We_Going. April 21, 2001.

Sununu, John. 2013. *Videos—John Sununu ICCC2.* The Heartland Institute. https://www.heartland.org/multimedia/videos/john-sununu-iccc2. July 5, 2013.

Sustainable. 2015. "Transforming Our World: the 2030 Agenda for Sustainable Development." SustainableDevelopment.un.org. 2015.

Svensmark, Henrik. 2009. "Cosmoclimatology: a new theory emerges." Wiley Online Library. https://onlinelibrary.wiley.com/doi/abs/10.1111/j.1468-4004.2007.48118.x. January 25, 2007.

TAT. 2017. *The Aspen Times.* "Aspen Mountain will open for skiing Memorial Day weekend." *Denver Post.* https://www.denverpost.com/?returnUrl=https://www.denverpost.com/ 2017/05/22/aspen-mountain-memorial-day-weekend-open/?clearUserState=true. May 22, 2017.

Taylor, James. 2011. "Climategate 2.0: New E-Mails Rock The Global Warming Debate." *Forbes.* Com. https://www.forbes.com/sites/jamestaylor/2011/11/23/climategate-2-0-new-e-mails-rock-the-global-warming-debate/#3342a0ad27ba. November 23, 2011.

—————.. 2018. "Election Slaughter for Climate Activism." American Thinker. https://www.americanthinker.com/articles/2018/11/election_slaughter_for_climate_activism.html. November 20, 2018.

Taylor, Kathryn. 1986. "Kathryn Taylor Weds T.F. Steyer." *New York Times.* https://www.nytimes.com/1986/08/17/style/kathryn-taylor-weds-tf-steyer.html. August 17, 1986.

Terra et Aqua. 2015. "Greenhouse gas solutions." Watertechbyrie.com. https://watertechbyrie.com/2015/05/02/greenhouse-gas-solutions/. May 2, 2015.

Thetruthpeddler. 2011. "Peak temperatures during Earth's present interglacial are cooler than during any interglacial of the past half million years." The Truth Peddler. https://thetruthpeddler.wordpress.com/2011/02/06/peak-temperatures-during-earths-present-interglacial-are-cooler-than-during-any-interglacial-of-the-past-half-million-years/. February 6, 2011.

Thomas, Andrew. 2014. "It's All About System Change (to Socialism), Not Climate Change." American Thinker. https://www.americanthinker.com/blog/2014/09/ its_all_about_system_change_to_socialism_not_climate_change.html. September 26, 2014.

Thompson, Bruce. 2010. "The Granularity of Climate Models." American Thinker. https://www.americanthinker.com/articles/2010/03/the_granularity_of_climate_mod.html. March 13, 2010.

Tijani, Mayowa. 2016. "Blocking us from coal-fired power is hypocrisy, Adeosun tells western powers." TheCable.ng. https://www.thecable.ng/blocking-us-from-coal-fired-power-is-hypocrisy-adeosun-tells-western-powers. October 5, 2016.

Toplanshy, Eileen F. 2018. "Population Stabilization or Suicidal Demographics?" American Thinker. https://www.americanthinker.com/articles/2018/10/population_stabilization_or_suicidal_demographics.html. October 23, 2018.

Tracinski, Robert. 2009. "ClimateGate: The Fix is In." Real Clear Politics. https://www.realclearpolitics.com/articles/2009/11/24/the_fix_is_in_99280.html. November 24, 2009.

—————. 2012. "Fakegate: Global Warmists Try to Hide Their Decline." https://www.realclearpolitics.com/articles/2012/02/23/fakegate_global_warmists_try_to_hide_their_decline_113225.html. RealClearPolitics.com. February 23, 2012.

—————. 2015. "Seven big failed environmentalist predictions." *The Federalist.* https://thefederalist.com/2015/04/24/seven-big-failed-environmentalist-predictions/. April 24, 2015.

Traufetter, Gerald. 2009. "Climatologists Baffled by Global Warming Time-Out." SpiegetOnline.com. http://www.spiegel.de/international/world/stagnating-temperatures-climatologists-baffled-by-global-warming-time-out-a-662092.html. November 19, 2009.

Trenberth, Kevin E., John T. Fasullo, et al. 2014. "Seasonal aspects of the recent pause in surface warming." Natural.com/Natural Climate change. https://www.nature.com/articles/nclimate2341. August 17, 2014.

Trinko. Tom. 2014. "Global Warming and the Feynman test." American Thinker. https://www.americanthinker.com/articles/2014/09/global_warming_and_the_feynman_test.html. September 11, 2014.

—————. 2018. "The Corruption of Science." American Thinker. https://www.americanthinker.com/articles/2018/06/the_corruption_of_science.html June 7, 2018.

Tsang, Derek. 2014. "Fewer wars, fewer people dying in wars now than in quite some time, Glenn Beck writer claims." Politifact.com. https://www.politifact.com/punditfact/statements/ 2014/jul/21/stu-burguiere/fewer-wars-fewer-people-dying-wars-now-quite-some/. July 21, 2014.

UN. 2017. "World population projected to reach 9.8 billion in 2050, and 11.2 billion in 2100." Un.org. https://www.un.org/development/desa/en/news/population/world-population-prospects-2017.html. June 17, 2017.

UCAR. 2008a. "Exosphere." UCAR Center for Science Education. https://scied.ucar.edu/shortcontent/exosphere-overview. 2008.

—————. 2008b. "Mesosphere." UCAR Center for Science Education. https://scied.ucar.edu/shortcontent/mesosphere-overview. 2008.

—————. 2008c. "Thermosphere." UCAR Center for Science Education. https://scied.ucar.edu/shortcontent/thermosphere-overview. 2008.

—————. 2011. Troposphere. UCAR Center for Science Education. https://scied.ucar.edu/shortcontent/troposphere-overview. 2011.

UDHR. 2019. "Universal Declaration of Human Rights." UN.org. http://www.un.org/en/universal-declaration-human-rights/index.html. 2019.

UEA. 2010. "Report of the International Panel set up by the University of East Anglia to examine the research of the Climatic Research Unit." Uea.ac.uk. http://www.uea.ac.uk/documents/3154295/ 7847337/SAP.pdf/a6f591fc-fc6e-4a70-9648-8b943d84782b. April 19, 2010.

UK Gov. 2008. Climate Change Act 2008. Legislation.goc.uk. http://www.legislation.gov.uk/ukpga/2008/27/contents. 2008.

UN. 2015. "World Population Prospects 2015." Esa.un.org. https://esa.un.org/unpd/wpp/publications/files/key_findings_wpp_2015.pdf. 2015.

Verkaik, Robert. 2009. "Just 96 months to save world, says Prince Charles." Independence. https://www.independent.co.uk/environment/green-living/just-96-months-to-save-world-says-prince-charles-1738049.html. July 9, 2009.

Vogt, William, Stuart I. Freeman, et al. 2007. *The Road to Survival.* Kissinger Publishing. https://www.goodreads.com/book/show/3304436-road-to-survival. August 23, 2007.

Vogt, William. The Road to Survival. William Sloan Associates. http://eps.berkeley.edu/people/lunaleopold/(014)%20Review%20of%20Road%20to%20Survival%20by%20W%20Vogt.pdf. 1948.

Walker, Bruce. 2008. "Madness and Political Life" American Thinker. https://www.americanthinker.com/articles/2008/04/madness_and_political_life.html. April 22, 2008.

Wallman, Brittany. 2018. "American Dream Miami, the nation's biggest mall, wins final OK." Sun.sentinal.com. May 17, 2018.

Watts, Adam. 2006. "24 days to Al Gore's '10 years to save the planet' and 'point of no return' planetary emergency deadline." Wattsupwiththat.com. https://wattsupwiththat.com/2016/01/02/24-days-to-al-gores-10-years-to-save-the-planet-and-point-of-no-return-planetary-emergency-deadline/. January 2, 2016.

Watts, Anthony. 2010a. "Professor Richmond Lindzen's Congressional Testimony." Wattsupwiththat.com. https://wattsupwiththat.com/2010/11/18/profess-richard-lindzens-congressional-testimony/. November, 18, 2010.

————. 2010b. "Willis publishes his thermostat hypothesis paper." Wattsupwiththat.com. https://wattsupwiththat.com/2010/07/24/willis-publishes-his-thermostat-hypothesis-paper/. Wattsupwiththat.com. July 24, 2010.

————. 2012. "NASA GISS caught changing past data again – violates Data Quality Act." Wattsupwiththat.com. https://wattsupwiththat.com/2012/09/26/nasa-giss-caught-changing-past-data-again-violates-data-quality-act/. 2012

————. 2014. "The big list of failed climate predictions." Wattsupwiththat.com. https://wattsupwiththat.com/2014/04/02/the-big-list-of-failed-climate-predictions/. April 2, 2014.

————.. 2015. "The Holocene Thermal Optimum." WattsUpWithThat.com. https://wattsupwiththat.com/2015/12/21/the-holocene-thermal-optimum/. December 21, 2015.

—————. 2016. "24 days to Al Gore's '10 years to save the planet' and 'point of no return' planetary emergency deadline." Whatsupwiththat.com. https://wattsupwiththat.com/2016/01/02/24-days-to-al-gores-10-years-to-save-the-planet-and-point-of-no-return-planetary-emergency-deadline/. January 2, 2016.

—————. 2017. "Defying Al Gore's predictions, bottom drops out of the US hurricane pattern over past decade." SOTT.net. https://www.sott.net/article/347530-Defying-Al-Gores-predictions-bottom-drops-out-of-US-hurricane-pattern-over-past-decade. April 6, 2017.

Watts, Nick, et al. 2017. "The Lancet Countdown on health and climate change: from 25 years of inaction to a global transformation for public health," Junkscience.com. https://junkscience.com/wp-content/uploads/2017/11/lancet-countdown-Oct-30-2017.pdf. October 30, 2017.

Wegman, Edward J. 2006. "Ad Hoc Committee Report in the 'Hockey Stick' Global Climate Reconstruction." Klimaatgek.nl. https://klimaatgek.nl/document/WegmanReport.pdf. 2006.

Weinstein, Adam. 2014. "Arrest Climate-Change Deniers." Gawker.com. https://gawker.com/arrest-climate-change-deniers-1553719888. March 28, 2014.

WFM. 2019. *Montreux document*. Wfm.igp.org. http://www.wfm-igp.org/our-movement/history. 2019.

Whatis. 2016. "chaos theory." Whatis.techtarget.com. https://whatis.techtarget.com/definition/chaos-theory. May 2016.

Whitehouse. 2018. "America First A Budget Blueprint to Make America Great Again." Office of Management and Budget, Page 41. https://www.whitehouse.gov/sites/whitehouse.gov/ files/omb/budget/fy2018/2018_blueprint.pdf. 2018.

Wikipedia. 100,000. 100,000-year problem. https://en.wikipedia.org/wiki/100,000-year_problem. 2019.

Wikipedia. AMO. Atlantic Multidecadal Oscillation. https://en.wikipedia.org/wiki/Atlantic_multidecadal_oscillation. 2019.

Wikipedia. ASR. "Ancestral sequence reconstruction." Wikipedia. https://en.wikipedia.org/wiki/Ancestral_sequence_reconstruction. 2019.

Wikipedia. Cambrian. Cambrian explosion, Wikipedia. https://en.wikipedia.org/wiki/Cambrian_explosion. 2019.

Wikipedia. Carboniferous. "Carboniferous." Wikipedia. https://en.wikipedia.org/wiki/Carboniferous. 2019.

Wikipedia. CCM. "Greenhouse gas emissions mitigation." Wikipedia. https://en.wikipedia.org/wiki/Climate_change_mitigation. 2019.

Wikipedia. CERN. Wikipedia - CERN, Wikipedia. https://en.wikipedia.org/wiki/CERN. 2019.

Wikipedia, Clarke. "Clarke's Three Laws." Wikipedia. https://en.wikipedia.org/wiki/Clarke%27s_three_laws. 2019.

Wikipedia. CPP. "Clean Power Plan." https://en.wikipedia.org/wiki/Clean_Power_Plan. 2019.

Wikipedia. Cretaceous. Cretaceous. https://en.wikipedia.org/wiki/Cretaceous. 2019.

Wikipedia. Cycles. "List of Solar Cycles." https://en.wikipedia.org/wiki/List_of_solar_cycles. 2019.

Wikipedia. Eocene. Eocene. https://en.wikipedia.org/wiki/Eocene. 2019.

Wikipedia. FYSP. Faint young sun paradox. https://en.wikipedia.org/wiki/Faint_young_Sun_paradox. 2019.

Wikipedia, Hayek. F.A. Hayek. https://en.wikipedia.org/wiki/Friedrich_Hayek. 2019.

Wikipedia. HL. Henry's Law. https://en.wikipedia.org/wiki/Henry%27s_law. 2019.

Wikipedia. Holocene. Holocene. https://en.wikipedia.org/wiki/Holocene. 2019.

Wikipedia. Jurassic. Jurassic. https://en.wikipedia.org/wiki/Jurassic. 2019.

Wikipedia. Kerry. 2019. "John Kerry." https://en.wikipedia.org/wiki/John_Kerry. 2019.

Wikipedia. LGM. Last Glacial Maximum. https://en.wikipedia.org/wiki/Last_Glacial_Maximum. 2019.

Wikipedia. MC. Milankovitch cycles. https://en.wikipedia.org/wiki/Milankovitch_cycles. 2019.

Wikipedia. MGR. Brunhes–Matuyama geomagnetic reversal. https://en.wikipedia.org/wiki/Brunhes–Matuyama_reversal. 2019.

Wikipedia. MM. Maunder Minimum. https://en.wikipedia.org/wiki/Maunder_Minimum. 2019.

Wikipedia. MP. Middle Pleistocene. https://en.wikipedia.org/wiki/Middle_Pleistocene. 2019.

Wikipedia. Neo. Neoproterozoic. https://en.wikipedia.org/wiki/Neoproterozoic. 2019.

Wikipedia. OF, Orbital Forcing. https://en.wikipedia.org/wiki/Orbital_forcing. 2Wikipedia. Ordovician. Ordovician. https://en.wikipedia.org/wiki/Ordovician. 2019.

Wikipedia. PA. Paris Agreement. https://en.wikipedia.org/wiki/Paris_ Agreement. 2019.

Wikipedia. PETM. Paleocene-Eocene Thermal Maximum. https://en.wikipedia.org/wiki/Paleocene–Eocene_Thermal_Maximum. 2019.

Wikipedia. PFOTC. Polar Forests of Cretaceous. https://en.wikipedia.org/wiki/Polar_forests_of_the_Cretaceous. 2019.

Wikipedia. Phanerozoic. Phanerozoic. https://en.wikipedia.org/wiki/Phanerozoic. 2019.

Wikipedia. Pleistocene. Pleistocene. https://en.wikipedia.org/wiki/Pleistocene. 2019.

Wikipedia. Ramanujan. Srinivasa Ramanujan. https://en.wikipedia.org/wiki/Srinivasa_Ramanujan. 2019.

Wikipedia. ROOCI. "Radius of outer closed isobar." https://en.wikipedia.org/wiki/Radius_of_outermost_closed_isobar. 2019.

Wikipedia. Silurian. Silurian. https://en.wikipedia.org/wiki/Silurian. 2019.

Wikipedia. Snowball. Snowball Earth. https://en.wikipedia.org/wiki/Snowball_Earth. 2019.

Wikipedia. Solar. Solar luminosity. https://en.wikipedia.org/wiki/Solar_luminosity. 2019.

Wikipedia. Sturtian. "Sturtian Glaciation." https://en.wikipedia.org/wiki/Sturtian_glaciation. 2019.

Wikipedia. Sun. Solar. https://en.wikipedia.org/wiki/Sun. 2019.

Wikipedia. Transhumanism. "Transhumanism." https://en.wikipedia.org/wiki/Transhumanism. 2019.

Wikipedia. UNFCCC. United Nations Framework Convention on Climate Change. https://en.wikipedia.org/wiki/United_Nations_Framework_Convention_on_Climate_Change. 2019.

Wikipedia. Vostok. Vostok Station. https://en.wikipedia.org/wiki/Vostok_Station#Ice_core_drilling. 2019.

Williamson, Kevin D. 2014. "Capitalism Is Green(er)." *National Review.* https://www.nationalreview.com/2014/09/capitalism-cleaner-kevin-d-williamson/. September 25, 2014.

WMO. 2006. *State of the Climate in 2006.* World Meteorological Organization. https://library.wmo.int/pmb_ged/wmo_1016_en.pdf. 2006.

Wojick, David E., and Patrick J. Michaels. 2015. "Is the Government Buying Science or Support? A Framework Analysis of Federal Funding-induced Biases." CATO Institute. https://www.cato.org/publications/

working-paper/government-buying-science-or-support-framework-analysis-federal-funding. April 30, 2015.

Wolfe, William L., and George J. Zissis. 1985. *The Infrared Handbook*. General Dynamics, revised edition. https://www.amazon.com/The-Infrared-Handbook-William-Wolfe/dp/096035901X. June 1, 1985.

Wright, Dexter. 2010a. "Climategate: How to Hide the Sun." American Thinker. https://www.americanthinker.com/articles/2010/01/climategate_how_to_hide_the_su.html. January 14, 2010.

—————. 2010b. "Just Sign on the Dotted Line." American Thinker. https://www.americanthinker.com/articles/2010/01/climategate_just_sign_on_the_d.html. January 24, 2010.

WTO. 1994. "The 128 countries that had signed GATT by 1994." World Trade Organization. https://www.wto.org/english/thewto_e/gattmem_e.htm. January 1, 1995.

Wulfsohn, Joseph A. 2019. "NBC's Chuck Todd defends not having climate change skeptics on 'Meet the Press' special" FoxNews.com. https://www.foxnews.com/entertainment/nbcs-chuck-todd-defends-not-having-climate-change-skeptics-on-meet-the-press-special. January 25, 2019.

Yoshida, Adam. 2011. "The Warmist's Dilemma." American Thinker. https://www.americanthinker.com/articles/2011/03/the_warmists_dilemma.html. March 4, 2011.

Young, Gregory. 2008. "Global Warming? Bring it On." American Thinker. https://www.americanthinker.com/articles/2008/11/global_warming_bring_it_on.html. November 21, 2008.

Zhu, Zaichun, Shilong Piao, et al. 2016. "Greening of the Earth and it drivers." Nature.com. https://www.nature.com/articles/nclimate3004. April 25, 2016.

Zombie. 2014. "Climate Movement Drops Mask, Admits Communist Agenda." PJ Media. https://pjmedia.com/zombie/2014/09/23/climate-movement-drops-mask-admits-communist-agenda/?singlepage=true. September 23, 2014.

Zubrin, Robert. 2013. Merchants of Despair, Encounter Books; reprint. https://www.amazon.com/Merchants-Despair-Environmentalists-Pseudo-Scientists-Antihumanism/dp/159403737X. December 31, 2013.

—————. 2017. "Merchants of Despair." Encounter Books. https://play.google.com/store/books/details?id=h7M_DwAAQBAJ&rdid=book-h7M_DwAAQBAJ&rdot=1&source=gbs_vpt_read&pcampaignid=-books_booksearch_viewport. November 21, 2017.

About the Author

AFTER RECEIVING a degree in Economics, L. Rowand Archer has enjoyed a 48-year career in the private sector where he successfully worked up the management ranks of various banking organizations, then founded a software development company which met the evolving business technology needs of the manufacturing sector, providing integrated administration and production controls making businesses more efficient and effective.

This work required a real-world understanding of the work people were undertaking and finding solutions that made sense to them in the long run.

Drawing from his extensive front line experience in multiple business operational environments, Mr. Archer is able to apply common sense observations of the uncertainty that his fellow citizens are experiencing with the rapid changes occurring in their daily life because of the new technologies being present to them. It is because of this uncertain environment that the population is vulnerable to the efforts of the liberal left and their partners in the liberal media to deliberately destabilize the common sense and critical thinking of these ordinary thinkers in an effort to advance a socialist agenda using a campaign of fear through climate alarmism.

After publishing the successful book, The Trump Effect, Unmasking the Dark-Side Left and Their Liberal Media Parrots, Mr. Archer felt it necessary to join the legions of authors attempting to expose the money-grubbing motives of unethical scientists, politicians, and charlatans promoting their manufactured climate change crisis based on questionable science and false global doomsday scare tactics. Mr. Archer wrote this book using easily accessible evidence to expose this climate warming fraud in the hope of stopping the alarmists in their real objective of advancing 'Progressive Socialism' and its subversion of our Free Market Institutions, which are based on individual freedoms, common sense, and reason.

Printed in the United States
By Bookmasters